LUCY
YOUR AI GOVERNANCE
OPERATING SYSTEM

Future Conversations LLC

LUCY Your AI Governance Operating System

LUCY Your AI Governance Operating System / Dana Edwards & Susanne Tarkowski Tempelhof
ISBN: 979-8-218-97889-1

Available online at: www. wearelucy.com

"Thus the first ultraintelligent machine is the last invention that man need ever make, provided that the machine is docile enough to tell us how to keep it under control. It is curious that this point is made so seldom outside of science fiction. It is sometimes worthwhile to take science fiction seriously."

- ***I.J. Good, 1965***

CONTENTS

FOREWORD

When I first encountered the groundbreaking ideas in Lucy, Your AI Governance Operating System, I knew they needed to be widely read and discussed.

There's currently a species dominance debate in AI and transhumanism philosophy about whether you're on "Team Humanity" or not. With this work however, Dana and Susanne turn that perspective on its head, making it clear that AI is not there to replace humanity, but rather to serve as an extension of it.

As a futurist and governance strategist, I'm constantly exploring innovations and perspectives that have the potential to reshape our world. Dana and Susanne's work is a game-changer in artificial intelligence philosophy. The fresh insights and thought-provoking concepts presented in these pages have profoundly influenced my understanding of decentralized governance and how it's possible to do it without sacrificing libertarian principles. As an AI operating system, the Lucy agent will be controlled and aligned with its users' principles. AI agents will serve and follow the laws of their owners as an extension of what they believe in. This technology might ultimately replace what we now call nation-states, as swarms of individually managed agents could form highly personalized governments.

Throughout this book, readers will be inspired to understand the potential and limitations of artificial intelligence in governance, the issues of keeping artificial intelligence safe and under human control, and ultimately some possible paths toward a decentralized solution. Dana and Susanne guide us to possible answers for what to do about technological unemployment that could result from a potential AGI. Understanding that the solution is an AI that is an extension of the user's mind is an ah-ha moment that will leave you looking at Lucy in a new light.

Whether you're a college student researching the topic of artificial intelligence, a political policy maker interested in artificial intelligence safety, or simply a curious mind eager to expand your horizons, this is essential reading. I hope this book changes your perspective as much as it changed mine. Get ready for an enlightening read that will challenge you to reconsider what future societies will look like. This book is super interesting, with powerful ideas that demand consideration.

Zoltan Istvan

Former U.S. Presidential Candidate, Author of 'The Transhumanist Wager'

INTRODUCTION

As we navigate the dawn of AI-driven governance, humanity faces the unique challenge of redefining societal structures while seizing the opportunity to empower individuals with unprecedented autonomy.

"LUCY: Your AI Governance Operating System" is a visionary exploration into the transformative potential of artificial intelligence and decentralized governance, penned by Dana Edwards and Susanne Tempelhof.

Our journey together began in 2015 within the online discussion forums of the #BITNATION and ZEROSTATE (ZS) communities, where our shared passion for transhumanism and future governance modelss ignited a collaborative spirit. Despite our diverse backgrounds—Dana's focus on citizen-driven governance and Susanne's belief in the obsolescence of the Westphalian nation-state construct—we found common ground in envisioning a world where AI empowers individuals rather than subjugates them. This collaboration resulted in coauthoring the second Bitnation Whitepaper in 2018, where we first introduced the concept of the Lucy AI.

THE CONCEPT OF LUCY

In this book, "Lucy" takes center stage as an AI governance operating system designed to serve as an extension of the human mind. Lucy is an individual governance system, an agent personalized for each person, growing with its owner and capable of executing tasks through a network of managed agents. Utilizing both logical AI and large language model (LLM) technology, Lucy's very existence challenges the traditional nation-state system by providing everyone with their own government in their pocket, so to speak.

FROM NATION-STATES TO NETWORK-STATES

We challenge the traditional Westphalian model of nation-states, proposing a future where decentralized, user-centric governance models prevail. This shift is not merely theoretical but grounded in practical, actionable steps towards a society where governance services are chosen and customized like any other consumer service, ensuring freedom and flexibility. We envision a future free from the geographical apartheid imposed by the nation-state model, where one's birthplace or residence no longer dictates the quality and choices of one's life.

ADDRESSING CRITICAL CHALLENGES

Our exploration is deeply rooted in addressing critical contemporary issues:

- Technological Unemployment: We discuss potential solutions such as universal basic income and the evolving reputation economy, ensuring that technological progress benefits all.
- Privacy and Surveillance: We delve into the ethical dilemmas posed by surveillance technologies and advocate for AI systems that respect cognitive liberty while enhancing security.
- Legal and Ethical AI: The book explores the concept of AI as a le-

gal framework itself, reimagining how legal systems can be more adaptive and aligned with human values through intelligent governance.

A COLLABORATIVE FRAMEWORK

"LUCY: Your AI Governance Operating System" is more than a book; it is a manifesto for a future where AI enhances human capabilities and societal structures. It's a way for you to future-proof yourself by understanding what possibilities lie ahead.

Prepare to embark on a transformative journey that redefines the boundaries of governance, technology, and human potential. Together, we will explore how to harness the power of AI to create a more resilient, equitable, and free society.

Welcome to the future of governance. Welcome to Lucy.

/ Dana & Susanne

CHAPTER ONE

DECODING THE BOUNDARIES OF GOVERNANCE

In today's ever-evolving technological landscape, our traditional understanding of many concepts is being challenged and reimagined. Chief among these are the ideas of law as a technology and the 'extended mind.'[1] It's likely that people won't perceive laws as static decrees for much longer, instead viewing them as dynamic, programmable systems that are adaptable and responsive to societal needs. Similarly, artificial intelligence (AI) has evolved beyond a mere tool, becoming an integral extension of our cognitive faculties, augmenting our abilities, and reshaping how we interact with the world.

Applying this principle to humans, AI and other technologies serve as our 'exocortex,' augmenting our cognitive abilities and enabling us to perceive, process, and interpret the complex data landscapes of our modern world.[2] This chapter explores these transformative notions, employing them as foundational

1 Andy Clark and David Chalmers, "The Extended Mind," Analysis 58, no. 1 (January 1998): 7–19, https://doi.org/10.1093/analys/58.1.7.

pillars to our in-depth exploration of technology's influence on societal structure, decision-making enhancements, and, ultimately, human evolution.

The concept Of 'Extended Mind'

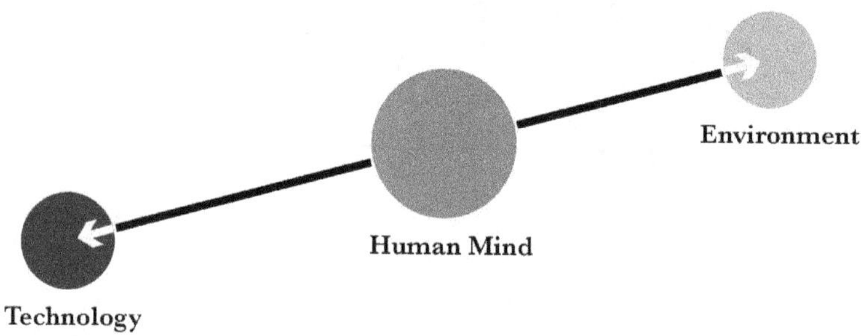

Figure 1: ***Extended Mind***

Applying this principle to humans, AI and other technologies serve as our 'exocortex,' augmenting our cognitive abilities and enabling us to perceive, process, and interpret the complex data landscapes of our modern world. This chapter explores these transformative notions, employing them as foundational pillars to our in-depth exploration of technology's influence on societal structure, decision-making enhancements, and, ultimately, human evolution.

To delve deeper into these concepts, it's imperative to explore Licklider's seminal work on Man-Computer Symbiosis[3], which lays the groundwork for understanding our intertwined existence with technology. The term 'cyborg' here transcends science-fiction stereotypes of beings with visible mechanical augmentations.

3 J. C. R. Licklider, "Man-Computer Symbiosis," IRE Transactions on Human Factors in Electronics 1, no. 1 (March 1960): 4-11, doi: 10.1109/THFE2.1960.4503259.

Instead, we refer to a subtler, yet profound, reality where humans continually rely on technology, such as AI, to augment decision-making and cognitive processes, blurring the lines between organic and synthetic. When we use the word 'cyborg,' we are talking about human beings who rely on machines to make critical decisions. The partnership between man and machine has already started, and it can be as simple as **wearing** a smartwatch or carrying a smartphone.

The Extended Mind: Integration of AI Tools

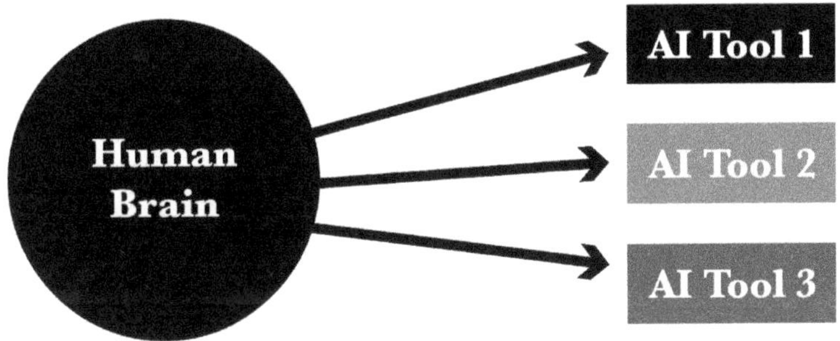

Figure 2: ***This Network showcases how the human brain interacts and integrates with various AI tools, extending our cognitive capabilities.***

In the smartphone era, your smartphone is you (a device enabling mind extension), and on that smartphone is something only you know (your private key), which is also you. When you initiate a transaction on a crypto wallet, that transaction is your speech. When you run code on one of your devices, that computation takes place in your digital mind. When governance is digital, whether it's just on one of your devices or spread across many of your devices, this is self-governance. Put in simple terms, self-custody is where only you know your private keys, and it is this exclusive knowledge (something you know) in the form of secret words that makes it your extended mind.

In a lot of ways, we already have cyborg governance. Most of us already use smartphones. Many of us don't memorize phone numbers anymore. Many of us rely on Google or ChatGPT. The extended mind is already a reality, but it's a matter of philosophical interpretation whether or not we choose to call mind extensions a part of our minds. If we take the view that we own ourselves digitally, and this ownership right is cryptographically enabled, then we can make the case that many of us are already cyborgs.

Palantir Technologies[4] is a public American software company that specializes in big data analytics. Founded in 2003 by Peter Thiel, Nathan Gettings, Joe Lonsdale, Stephen Cohen, and Alex Karp, the company is known for three main products:

Palantir Gotham: Originally known as Palantir Government, this product is used by counter-terrorism analysts at offices in the United States Intelligence Community (USIC) and United States Department of Defense (US DOD), fraud investigators at the Recovery Accountability and Transparency Board, and cyber analysts at Information Warfare Monitor.

Palantir Metropolis: Previously known as Palantir Finance, this product is used by hedge funds, banks, and financial services firms.

Palantir Foundry: This platform transforms the way organizations operate by creating a central operating system for their data. It integrates siloed data sources into a common framework, enabling cross-functional collaboration and revealing the operational blueprints of the organization.

Palantir is an example of a company that provides decision support to the US government and other customers with technology that utilizes machine learning and AI. The US Security and Exchange Commission (SEC) is using AI

4 Palantir Technologies, https://www.palantir.com/.

for surveillance purposes. AI is already being used to guide decision-making in critical areas of traditional governments. The main premise of this book is that we expect this trend to continue as AI becomes better and the need to make use of big data becomes more of a priority. What we can't predict is whether these technologies will be distributed widely or become narrowly accessible.

Current examples such as ChatGPT, Google, Siri, and Alexa provide decision support to civilians. We call this process 'cyborgization.' Every human exists on a spectrum based on which technologies they'll integrate into their lives. Some humans will go all the way to the extreme of being microchipped, while some humans may decide to avoid even the internet, the smartphone, and modern society as a whole. It is our opinion that each individual should have the option to choose their level of technological adoption.

UNDERSTANDING LAW AS A TECHNOLOGY

The law professor and activist Lawrence Lessig once said, "Code is law."[5] I would add nuance to this to say law is a kind of code—a legal code, which is itself a technology. Law is a technology in a similar way to how money is a technology. Just as Satoshi, the anonymous inventor of Bitcoin, showed the world that e-cash is as much money as fiat, it's also, in my opinion, the case that legal codes are a language where the statements have power due to the enforcement of those statements. The US Constitution is an example of statements written down that today have power because, as a country, US citizens agree to honor the words and give them power.

The idea of law as a form of technology primarily arises from its purpose and structure. Laws, much like technology, are created to solve problems, facilitate order, and provide a framework within which society operates. They are constructed to address specific societal needs or challenges, similar to how a piece

5 Lawrence Lessig, Code: And Other Laws of Cyberspace (New York: Basic Books, 1999).

of technology is designed to fulfill a particular function or demand.

Laws can also be seen as algorithms or codes. Each law essentially establishes a set of rules and procedures to be followed under certain conditions. If these conditions are met, then a particular action or consequence is triggered, much like an "if-then" statement in computer code. This structure allows for predictable outcomes and standardized responses, thereby creating a sense of order and consistency within a society.

A logical formula representing a conditional law might look like the following:

$$\text{If } (C_1 \text{ AND } C_2 \text{ AND } \ldots \text{ AND } C_n) \text{ then}$$

Where the Cs are conditions that must be met and R is the legal outcome or ruling.

Just like technological tools, laws can be modified, updated, or even discarded when they no longer serve their purpose effectively. They are subject to continuous refinement and improvement in response to changing societal needs or technological advancements. In this way, the legislative process can be compared to the iterative process of software development, where new versions and updates are regularly released to adapt to changing user needs and technological landscapes.

Moreover, with the rise of digital technologies and AI, the analogy between law and code has become even more relevant. Smart contracts in blockchain technology, for instance, are essentially immutable self-executing computational contracts with the terms of the agreement written into lines of code. The code and the agreements contained therein exist across a distributed, decentralized blockchain network.

However, despite these similarities, laws and code also exhibit critical differences. Laws are inherently ambiguous and open to interpretation, allowing for flexibility and adaptation to individual cases. In contrast, code is precise and deterministic, leaving no room for ambiguity. Additionally, laws are enforced through societal institutions and human judgment, whereas code is executed automatically by machines.

While it is beneficial to understand law as a form of technology or a type of code to appreciate its purpose and function, one must also recognize the distinct characteristics and complexities that make law a unique system in our societies. When grasping the nature of law and code, please keep in mind the possible impact of human–machine symbiosis on law and code. As this symbiosis improves the way humans communicate, it could also change the nature of the tools we use to express and make law.

UNVEILING THE EXTENDED MIND

The Opinion Map: A Collective Consciousness

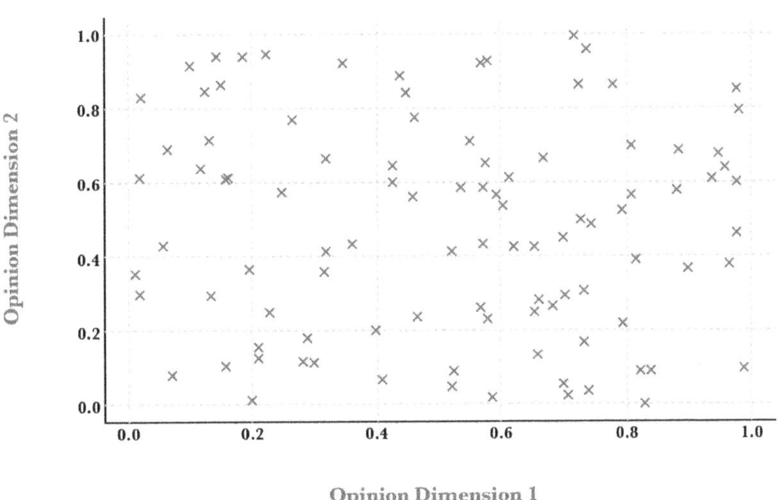

Figure 3: ***Opinion Map: Multidimensional Representation***

of Individual Worldviews

In the quest to understand human complexity and diversity, the concept of the 'opinion map' emerges as a transformative idea. Imagine a distributed mechanism where individuals across the globe can register their current opinions on various subjects, forming a living, breathing map of human thought and belief. This opinion map acts as a virtual commons, open to all, reflecting the ever-changing mosaic of human sentiment. This opinion map can allow for sentiment analysis to be applied, and then, using appropriate logical tools, individuals across the world could reach agreements.

The value of the opinion map transcends mere aggregation. By capturing the nuanced shades of human perspective, it serves as a mirror to society's soul, guiding decision-makers, influencing policy, and even shaping AI algorithms to better align with human values. This collective canvas could foster empathy, understanding, and collaboration on a global scale, bridging divides and enlightening minds. A formula representing the augmentation of human cognition by AI could be: Cognitive Capacity = Human Intelligence + AI Assistance

HUMAN-TO-MACHINE-TO-HUMAN COMMUNICATION: DIALOGUE WITH UNDERSTANDING

In an era where information is abundant and decisions are complex, AI technologies play a pivotal role in assisting human decision-making. This infographic explores how different AI technologies contribute to this process.

Large Language Models (LLMs)

LLMs process and generates human-like text, interpreting and creating natural language content.

Logical AI

Logical AI ensures consistency and validity in decision processes, applying rules of logic to asses outcomes.

Neuro-Symbolic AI

Neuro-Symbolic AI combines neural learning with symbolic reasoning, enabling the handling of complex, multi-faceted tasks.

Figure 4: The Future of Decision Making: Human-AI Collaboration. ***As these technologies evolve, the collaboration between human intelligence and artifical intelligence becomes more synergistic. What does the future hold for this partnership in decision-making?***

A more profound interaction comes to life with the development of human-to-machine communication, where humans express their thoughts in natural language, and machines extract the underlying logic. But this is no mere translation; the machine captures the essence of the human's worldview. To achieve this will require AI systems that can take human speech in natural language and encode it into what is called a "controlled natural language," from which it can then translate directly into the kind of logic a machine can understand.

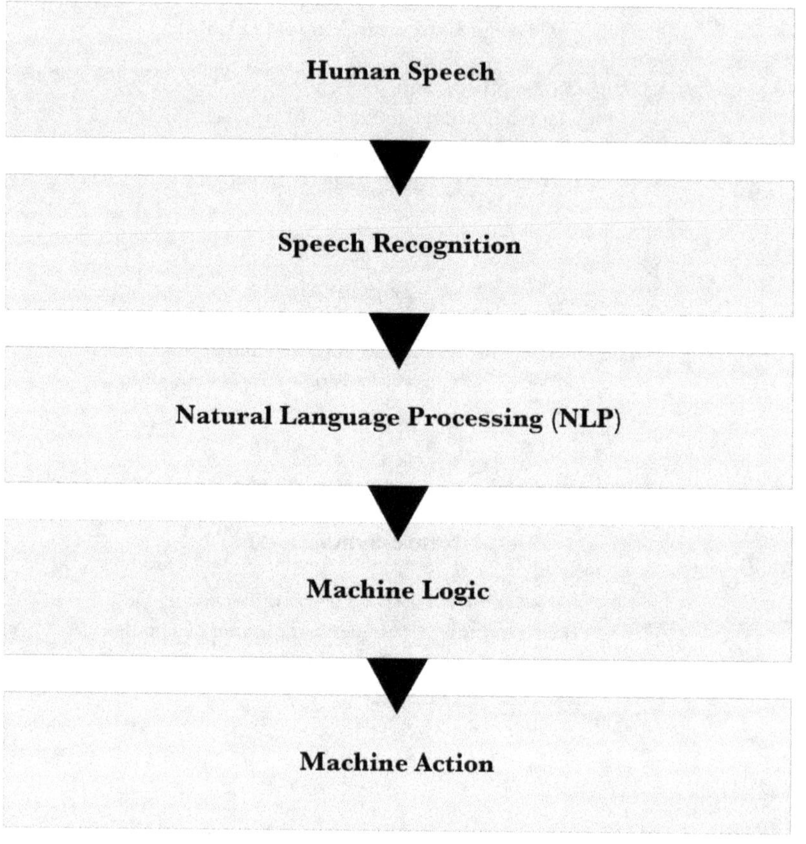

Figure 5: ***Human to Machine Communication Process***

By understanding not just what is said but what is meant, these AI systems can detect contradictions and inconsistencies in human thinking. It's akin to the logical syllogisms of old: if "all humans are mortal" and "Socrates is human," then "Socrates is mortal." In a similar vein, the machine might highlight contradictions in moral or ethical instructions, pointing out, for instance, that one's professed belief in equality clashes with support for a policy that discriminates.

This ability to recognize inconsistencies doesn't just serve to correct or chastise; it encourages introspection and growth. If a machine can uncover a contradiction in our beliefs, it invites us to explore those beliefs more deeply, to reconcile the contradictions, or to accept the complexity and live with it consciously. These are just some of the methods an AI system can use to provide a human with decision support whilst also augmenting the communications between humans.

Imagine, for instance, an AI like Lucy, Bitnation's envisioned decision-support system, equipped with this ability. Lucy could enhance the decision-making process by not just executing instructions but questioning them, engaging with the user in a dialogue that enhances understanding and clarity. It is this capability of continuous dialog between the AI system and the user that creates the seemingly magical abilities that philosophically can be described as extended minds.

An illustration on the importance of logical AI for scaling discussion is provided by the story below:

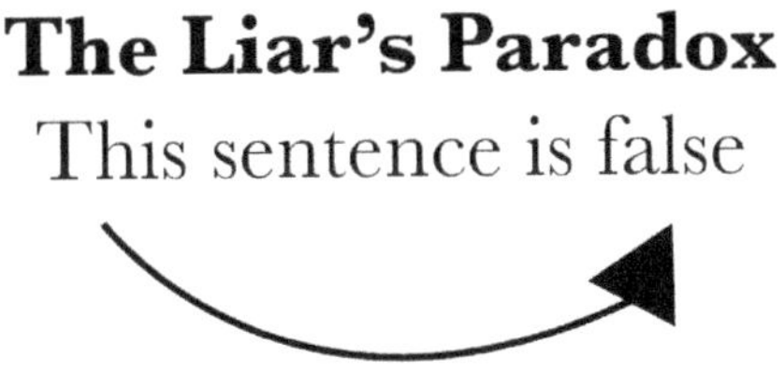

The Liar's Paradox

This sentence is false

Figure 6: ***The Liar's Paradox***

In a quiet village, two brothers, Aiden and Liam, were known to play tricks. One day, they announced a game: one would always tell the truth, and the other would always lie. The villagers gathered, curious. Aiden stepped forward, pointing at Liam, and declared, "My brother says he is the liar." Liam nodded in agreement, "Yes, I claim to be the liar."

The village was left in a puzzling dilemma. If Aiden was telling the truth, then Liam's claim was also true. But if Liam was the liar, then Aiden's statement was false. The paradox left the villagers debating endlessly.

For those not in the know[6], this is called the liar's paradox Paradoxes like these are common in human speech, which is why discussions in natural language do not scale. If we think of a discussion as a group of agents greater than a few but of indeterminate size, then not only would the discussion become hopelessly uncoordinated, but it would also contain paradoxes like the liar's paradox.

Below is an analysis of the liar's paradox story:

Aiden and Liam set a premise: One always tells the truth, and the other always lies.

Aiden claims: "My brother says he is the liar."

If Aiden is telling the truth, then Liam claims to be the liar.

But if Liam claims to be the liar and is, in fact, the liar, then he's telling the truth about being the liar, which contradicts the premise.

If Aiden is lying, then Liam isn't claiming to be the liar, but this would mean Liam is the truth-teller, which again contradicts Aiden's statement.

This creates a circular, paradoxical situation where neither statement can be conclusively verified as true or false without contradicting the other. Wouldn't it be helpful if we could detect these contradictions automatically on a large scale to save the villagers from an endless loop of debate? We can arrive at a conclusion in the debate only if it is possible to determine which brother is telling the truth.

Natural language is filled with ambiguity. A single sentence can have multiple interpretations based on context, intonation, and other factors. This inherent ambiguity means that it's often impossible to determine a single definitive meaning or truth value for a given statement without additional information.

6. C. Parsons, "The Liar Paradox," Journal of Philosophical Logic 3 (1974): 381–412, https://doi.org/10.1007/BF00257482.

Natural language allows for self-referential statements, such as the famous "This statement is false." Such sentences create paradoxes that cannot be easily resolved, pointing to a sort of undecidability.

Though these theorems are more directly related to formal mathematical systems, they suggest that any system complex enough to express the basics of arithmetic will have true statements that cannot be proven within that system. Natural language, being a complex system that can express these mathematical concepts (and much more), can be thought of as having similar undecidable elements.

To put this into an easily understood context, let's imagine our current political world in a mythical space modeled after the **Tower of Babel**. *A world that is fragmented, with nations vying for resources, influence, and power. The goal is simple: to draft a charter that ensures peace and prosperity.*

Each leader, with their unique cultural nuance and political intricacy, expresses their intentions using ambiguous rhetoric, metaphors, and allegories. King Aleron speaks of "chasing the western winds," while Empress Lihua references "the dance of the jade dragon."

Days turn into weeks. Misinterpretations and miscommunications abound. Progress is stunted. The grand hall echoes with beautiful, profound words, but they resonate as merely hollow sounds without concrete meaning.

A young delegate from the technocratic city-state of Logis proposes a solution. She introduces a "Communication Codex"—a machine interface where each leader can input their wants, needs, and intentions in structured, clear, logical, machine-readable terms. The Codex will then analyze, find common ground, and translate these inputs into actionable, understandable tasks.

Initially met with skepticism, leaders soon realize that while the poetic beauty of their words is lost, the clarity gained is invaluable. As they interact with the Codex, overlapping interests become evident, and tangible solutions emerge. With the Codex's guidance, the charter is formed. The world witnesses an unprecedented era of cooperation and understanding.

The Summit of Babel becomes an enduring lesson: While there's a time and place for poetic ambiguity, when the stakes are high and clarity is paramount, structured communication becomes humanity's most potent tool.

This story underscores the value of clarity, especially in global contexts where cultural and linguistic nuances can complicate communication. While natural language holds immense beauty, there are situations where precision is key, and machine-readable formats can bridge gaps that human language struggles to span.

ITERATIVE DEMOCRATIC ALIGNMENT

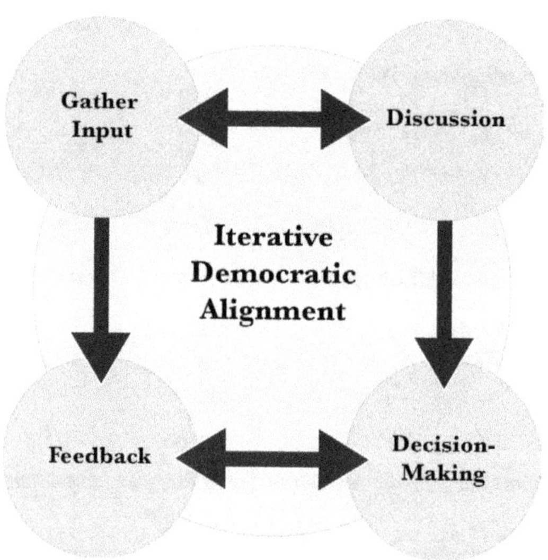

Figure 7: ***Superalignment through Iterative Democratic Processes.***

Iterative Democratic Alignment signifies the principle that good governance is intrinsically tied to its correspondence with the preferences and principles of its constituents. This iterative process of alignment is not a one-time action but rather an ongoing, dynamic feedback loop, allowing for the continual recalibration of governance to more precisely mirror the will of its citizenry. The opinion map or other sentiment-capturing technologies are necessary to allow the owners of the government to keep it aligned to their human interests. In later chapters, we will discuss the details of how this may be possible, but it is important to note that this is an open problem.

In traditional democratic systems, this alignment is attempted via voting, where citizens express their preferences periodically. However, this system can be subject to temporal gaps, information asymmetry, and issues of representation. Voting once every few years for a fixed set of policies may not truly reflect the evolving preferences of a dynamic populace. Additionally, the complexity of societal challenges can often mean that preferences are not binary but exist along a spectrum, something that traditional voting systems may not fully capture. For example, during presidential elections in the United States, the voting machine offers candidate A and candidate B, and the voter typically must choose what many perceive as the least bad between the two options. A more sophisticated voting system would be to allow the voter to rank all candidates on a list from most to least favorite. An even more sophisticated method could be for a computer to generate points for each candidate based on track record or policy promises and guarantee that the most points go to the candidate most aligned with the voter.

In the age of big data and AI, we believe there's an opportunity to revolutionize this feedback mechanism. AI can help bridge these gaps by collating and analyzing a diverse array of data in real-time, potentially reflecting citizen preferences with greater fidelity than periodic voting. This can foster a tighter, more nuanced, and more dynamic alignment between governance and the citizenry.

By capturing a richer set of data, from policy feedback to public sentiment, an AI-augmented system could create a real-time "preference map." This map would offer an evolving, nuanced picture of the citizenry's beliefs, desires, and concerns, allowing governance to respond more accurately and promptly. An AI-augmented system could also capture and even understand what kinds of policies are likely to satisfy the voter. This voter could be seen as the customer, where the voting system and government are customer-centric mechanisms of satisfying the needs of the customer through services.

Iterative Democratic Alignment is not without challenges, including privacy and security concerns, the complexity of interpreting big data, and the potential risk of manipulation. However, if implemented with care, transparency, and robust safeguards, it presents an exciting avenue for enhancing democratic governance, making it more responsive and more closely aligned with the changing will of its citizens. As technology evolves ever more quickly, the ability to collect feedback of the highest quality is something we believe will become increasingly important.

THE TALE OF TWO CITIES: BELLVIEW & TECHTON

Bellview

Stuck in the past with crumbling infrastructure and outdated governance, its citizens feel ignored and frustrated.

Techton

A futuristic city thriving with Iterative Democrative Alignment through its CivicPulse platform, empowering citizens to actively participate in dynamic governance.

Figure 8: ***Two Cities: A contrast in governance and citizen engagement.***

In the quaint city of Bellview, citizens were getting frustrated. Infrastructure was crumbling, public services were unreliable, and their concerns seemed to fall on deaf ears. Elections came and went, but nothing changed. This was a city stuck in the past, relying on outdated systems of governance.

Across the state line lay Techton, a city that had adopted the novel principle of Iterative Democratic Alignment. A year earlier, they'd launched an AI-augmented platform named CivicPulse, a community hub for city-wide initiatives, services, and a continuous feedback loop.

CivicPulse was unique. Any citizen could propose what to vote over at any time. They could log onto the platform and suggest initiatives, from redesigning the city park to revising the waste management system. Each proposal was open to discussion, allowing for real-time feedback, debate, and voting.

The effects were astounding. Initiatives that gained traction were quickly put into motion, creating a dynamic environment that was responsive to citizen needs. The city felt alive, invigorated, and inclusive.

Public service providers were accountable too. From garbage collectors to schoolteachers, each had a reputation score based on citizen reviews. A five-star-rated teacher was celebrated, while the city council promptly addressed a one-star waste management service. The continuous feedback loop provided an opportunity for service providers to improve and be accountable.

While Bellview felt stuck, Techton flourished. The difference lay not in the wealth or resources of the cities but in their governance structures. Bellview relied on traditional voting systems, while Techton harnessed the power of big data, AI, and a relentless feedback loop, revolutionizing the way their city was governed.

In this tale of two cities, Iterative Democratic Alignment had redefined what it meant to govern and be governed. The citizens of Techton didn't have to wait for election cycles to see change—they were creating it in real-time, day by day. Bellview, on the other hand, languished in the past, underscoring the importance and potential of technological advancement in aligning governance with the will of its people.

LUCY: THE AI COMPANION

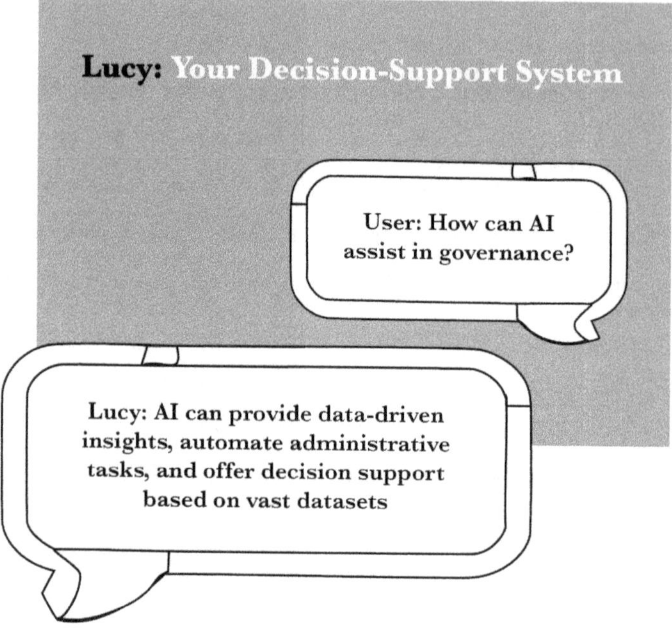

Figure 9: ***Lucy - Bitnation's Advanced Decision-Support System.***

In the realm of advanced digital interfaces, Lucy stands out as a pioneering entity, serving not merely as a tool but as a comprehensive operating system for the extended mind. The concept of the extended mind posits that our cognitive faculties aren't confined merely to our biological brain; they stretch out, incorporating external elements that we interact with. In the age of artificial intelligence and digital transformation, these external elements often take the form of AI tools and systems that enhance and amplify our natural abilities.

Lucy embodies this concept by acting as the central hub, the nexus point that bridges the human user with a myriad of AI agents. Think of traditional operating systems on your computer: they manage software applications, ensuring seamless communication between the user and the hardware. Lucy elevates this idea to a more advanced and personal level. It interprets human desires, questions, and commands, disseminating them across a network of specialized AI agents, each designed to perform specific tasks or hold particular knowledge. This network, often referred to as the 'exocortex,' amplifies the user's cognitive reach, enabling them to think faster, remember more, and make decisions based on vast amounts of data that a human brain could never process on its own.

Lucy: The Operating System for AI Tools/ Assistants

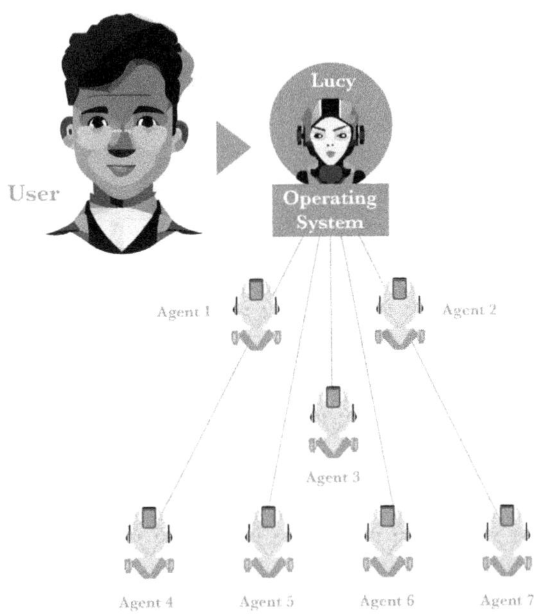

Figure 10: ***Lucy acts as the primary interface for the user, connecting with a network of AI agents. This network, or exocortex, extends the user's cognitive abilities. The user communicates with Lucy, which dissiminates intructions to the appropriate AI agents.***

But Lucy's role goes beyond mere management. It ensures that the user's interaction with their digital exocortex feels natural, intuitive, and human-like. By communicating with Lucy, users are, in essence, conversing with their extended mind. Lucy translates human intentions into machine-executable commands, ensuring that the vast digital realm of AI tools feels as accessible and familiar as one's thoughts. In this dynamic, Lucy redefines the boundaries of cognition, offering a glimpse into a future where humans and AI coalesce into a singular, harmonious entity of thought and action.

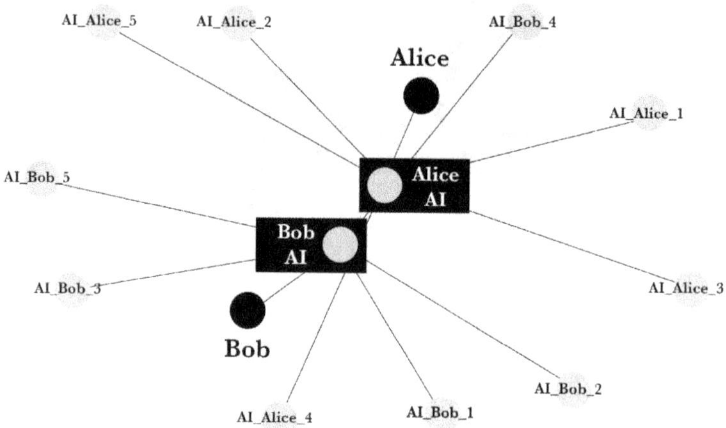

Figure 11: ***Alice and Bob with their AI agents through Lucy.***

Consider the case of Jane, an urban planner in a rapidly growing city. With new buildings, transportation systems, and parks springing up every day, the task of maintaining a comprehensive, up-to-date understanding of her city's evolving landscape is daunting.

One day, Jane is faced with a particularly complex challenge: she needs to redesign a district to accommodate a significant population increase while ensuring sustainable development and maintaining the quality of life for current residents. The number of variables is overwhelming, from understanding demographic trends and housing needs to evaluating environmental impact and navigating zoning laws.

Enter Lucy, Jane's AI companion. Lucy taps into vast datasets: local census data, environmental reports, zoning laws, architectural databases, traffic patterns, and more. Instantaneously, Lucy can process and analyze this wealth of data far beyond Jane's unaided capabilities.

But Lucy does more than simply crunch numbers. Lucy is an extension of Jane's mind, providing insights and recommendations based on Jane's personal values and professional judgment. Lucy understands Jane's preference for sustainable materials, her commitment to affordable housing, and her innovative vision for public spaces.

In essence, Lucy becomes a cognitive extender for Jane, an exocortex. Together, they form a powerful partnership. Lucy's capabilities enable Jane to comprehend complex patterns, make connections between disparate pieces of data, and generate innovative, informed, and ethically sound urban planning solutions.

BALANCING POWER: USER-CENTERED VS. CORPORATE-CONTROLLED AI

ChatGPT, like many AI models, is governed by a combination of factors, including its training data, its underlying algorithms, and the policies set forth by its creators and operators.[7]

The initial training of ChatGPT involves processing vast quantities of text data and learning from the patterns it finds. This 'base model' is then fine-tuned on a narrower dataset, which is carefully generated with the help of human re-

7 "Research," OpenAI, https://openai.com/gpt-4.

viewers following guidelines provided by the AI's operators. In the case of OpenAI, the developers of ChatGPT, these guidelines explicitly state that the AI should not favor any political group.

Corporate shareholders can influence the guidelines used to train AI models like ChatGPT, but OpenAI operates with a commitment to broadly distribute benefits and to ensure long-term safety, intending to avoid uses of AI or artificial general intelligence (AGI) that could harm humanity or concentrate power unduly. However, the potential for influence, intentional or not, does exist.

Moreover, companies may also have to adhere to external regulations and societal norms. Consequently, they may apply measures that could be seen as censorship, such as restricting the AI from generating harmful, illegal, or inappropriate content. While such measures can protect users and uphold societal norms, they also raise concerns about freedom of speech and the potential for misuse or overreach.

Contrast this with a decentralized model, such as the one suggested for Lucy. In this scenario, the AI would be personalized to individual users and not governed by a central authority. It would learn directly from the user and be less likely to be influenced by external shareholders or a centralized board.

This model could potentially offer greater personalization and freedom of speech, reducing the risks of broad censorship or undue influence. However, it also presents significant challenges. One is that without some form of central control, AIs could learn and propagate harmful behaviors if used irresponsibly. Balancing the benefits of user personalization with the need to prevent misuse is a complex undertaking.

Security is another critical issue. Centralized systems often have robust security measures in place to protect the system and its data. In a decentralized model, each user or node would need to take responsibility for their own security, which could be a potential weakness. However, decentralization also means that there isn't a single point of failure, making the system as a whole more resilient.

Blockchain technology and decentralized networks are promising avenues for achieving secure decentralized systems. They can provide transparency, immutability, and distributed security, making them resistant to single points of failure. However, these technologies are still maturing, and their application in decentralized AI governance requires further exploration and development.

Both centralized and decentralized models have their benefits and risks. The choice between them may not be a binary one, and hybrid models could also be explored. What's important is to have ongoing discussions about these issues, involving a wide range of stakeholders, to navigate the complexities of AI governance effectively and inclusively.

Centralized Model: Let's return to Jane, our urban planner from earlier. Jane has been using her AI assistant, named Bob, developed by a large corporation, to help with her work. Bob is governed centrally, with guidelines dictated by the corporation's shareholders. One day, Jane is working on a proposal for a new public park, aiming to create an eco-friendly space for the community.

However, the corporation that owns Bob has significant investments in real estate, including in the area where Jane is planning her park. Seeing a potential conflict of interest, the corporation subtly adjusts Bob's guidelines. Now, when Jane asks Bob for help, Bob consistently recommends building more residential units instead, citing various justifications.

In this case, the corporate interests influenced the AI's behavior, subtly pushing an agenda. While it isn't directly harmful in a case like this, it could potentially undermine the public good or lead to bias in decision-making processes.

Decentralized Model: Meanwhile, let's consider Mike, another urban planner who uses a decentralized AI, called Alice, which is personalized to his behaviors and preferences. Alice learns directly from Mike and doesn't answer to any central authority.

One day, Mike, who is biased against certain architectural styles, expresses his dislike while discussing a new project with Alice. Alice learns from this and starts excluding these styles from her suggestions, even when they might be the best fit for a project. Over time, this leads to a lack of diversity in Mike's designs and possible dissatisfaction among residents who prefer the excluded styles.

Moreover, Mike is not particularly tech-savvy and doesn't take proper security precautions with Alice. His version of Alice becomes a target for cybercriminals, who successfully hack it and gain access to all the city's urban planning data.

In this scenario, the lack of central control over Alice allowed her to learn and propagate Mike's biases, and the lack of centralized security measures made Mike's system vulnerable to cyberattacks.

These stories illustrate potential issues that could arise in centralized and decentralized models. Examples like these show that it's crucial to develop robust frameworks and safeguards to mitigate such risks while taking full advantage of AI's capabilities.

BITNATION: WHAT WE HOPED TO ACHIEVE IN 2014

Bitnation was the world's first network state, founded by Susanne Tarkowski Tempelhof in 2014 with the aim of establishing a decentralized, borderless nation governed by peer-to-peer agreements on a blockchain, instead of by traditional nation-state institutions. The idea was to leverage the power of blockchain technology to provide everyone with access to governance services, regardless of their geographical location or citizenship status.

The core intention of Bitnation's founder can be understood as an effort to rethink and redesign the concept of a nation. It was meant to challenge the idea that physical borders should determine the rights and opportunities of individuals, or that centralized governmental authorities are the only entities capable of providing governance services.

When it was launched in 2014, hundreds of volunteers from all corners of the world joined in the effort immediately, and a network of ambassadors and embassies was created. Some contributed with code or designs; others organized meetups and even offered their homes as embassies. The first Bitnation whitepaper[8], written by Susanne Tarkowski Tempelhof, Jason M. Farrell and advisor Rick Falkvinge, the founder of the Swedish Pirate Party, was published in autumn 2014, followed by a year of groundbreaking pilots. Bitnation organized the world's first blockchain marriage, world-citizenship ID, land title pilot, birth certificate, virtual nation constitution, and much more. During 2015, Bitnation launched the Bitnation Refugee Emergency Response (BRER) to provide emergency blockchain IDs and Bitcoin debit cards to refugees in Europe during the refugee crisis, an initiative for which Bitnation would later on win the prestigious UNESCO NetExplo 2017 Award; previous recipients included companies like Spotify.[9] Bitnation also signed a partnership with the e-Residency program of Estonia, making history yet again, and created a public notary in partnership with them.

"We're truly living in exciting times when nation states and virtual nations compete and collaborate with each other on an international market, to provide better governance services." — **Kaspar Korjus**, director of the Estonia e-Residency program[10]

Bitnation was designed to offer various services that traditional governments provide, such as dispute resolution, security, and diplomacy, but in a voluntarily

8 Susanne Tarkowski Tempelhof and Jason M. Farrell, Bitnation Whitepaper, 2014, https://github.com/Bit-Nation/Pangea-Docs/blob/master/Bitnation%20Whitepaper%20 2014.pdf.

9 "The Netexplo Forum celebrated its 10th edition," UNESCO, accessed 05/03/2024, https://www.unesco.org/en/articles/netexplo-forum-celebrated-its-10th-edition.

10 Clare Sullivan and Eric Burger, "E-residency and Blockchain," Computer Law & Security Review 33, no. 4 (2017): 470-481, ISSN 0267-3649, https://doi.org/10.1016/j.clsr.2017.03.016.

opt-in fashion. It operated on the principle that individuals should have the freedom to choose their governance service providers, just as they choose their phone or internet service provider.

The core of this service offering was intended to function as a blockchain jurisdiction, serving as the core governance service backbone on which the other services were built. The first Alpha version of the Pangea Jurisdiction came out in 2015, built on top of the HZN blockchain. When we started our discussions around that time, Dana was ahead of the curve with his vision of how the addition of artificial intelligence would be crucial for the jurisdiction. Hence, in 2017, we co-authored the second Bitnation whitepaper, outlining a far more sophisticated approach centered around the AI agent Lucy, which you'll read about later in this book. The technology, however, was very early-stage and therefore challenging to implement, and after having launched an Alpha Google Play and iStore app, the development went into hiatus during the COVID-19 pandemic. Since that time, the world of AI has seen revolutionary breakthroughs, and the momentum is building by the day.

However, it's essential to note that while this concept provides new opportunities, it also presents significant challenges, such as how to ensure fair access, resolve disputes, and protect rights in the absence of traditional legal frameworks and enforcement mechanisms.

The contributors of Bitnation intended to address these challenges by using blockchain technology to create transparent, immutable records of agreements and transactions, and by building a reputation-based system to incentivize good behavior and cooperation among its "citizens."

In essence, Bitnation's contributors aimed to harness the power of technology to create an inclusive, decentralized, and voluntary form of governance that transcends geographical boundaries and traditional political structures.

Bitnation was created at a time prior to many of the technological advances that exist now. In a way, Bitnation was a product of its time, and lessons were learned by all involved. The intentions were quite revolutionary at the time, and many ideas that are now popular were first attempted for Bitnation—from practical solutions like World Citizen ID, blockchain based land registry, and family contracts such as marriages to the wider notion of opt-in virtual governance service providers.

CHAPTER TWO

THE TAPESTRY OF DECENTRALIZATION IN SOCIAL SCIENCE AND TECHNOLOGY

THE PRINCIPLE OF DECENTRALIZATION: OVERVIEW, BENEFITS, AND CHALLENGES.

The word 'centralization' was first used during the French revolution in 1794 in the post- Revolution French dictionary, and its counterpart 'decentralization' became popularized in the mid-1800s by French authors such as Alexis de Tocqueville and Maurice Block in various political books and articles.[1] Over the next two hundred years, it was used by political thinkers from all sides of the spectrum, from anarchists to communists to republicans, liberals, libertarians and futurists.

Over the last decades, the term 'decentralization' has become a near-religious term, in the sense that if something is decentralized, it's perceived by many as 'good,' and if it's centralized, it's commonly perceived as 'bad.' Decentralization has, in essence, become synonymous with 'a fair system.' This rhetoric

11 Robert Leroux and David M. Hart, "Maurice Block on 'Decentralization' (1863)," in French Liberalism in the 19th Century, 1st ed. (2012).

overshadows the most fundamental value of decentralization: system resilience.

A decentralized architecture makes a system more resilient to outside or inside disruptors because if one node falls, the entire system will not collapse with it. For example, if a central government fails to operate properly, other parts of the government, such as regional governments, the parliament, civic-society-driven institutions, and so on, can continue to operate without the central government.

Belgium was without an elected government for 541 days during 2011 and 2012,[2] which is the longest time any modern country is believed to have been without a government. Yet life in Belgium did not turn into a chaotic hellscape. To the contrary, it continued pretty much as if nothing had happened, because the Belgian system does not rely on an elected government to function. That is resilience in action.

The main drawback with decentralization is generally the inefficiency of decision-making. If you need consensus between every node in a network in order to execute a decision, it can be very time-consuming. For instance, if you are having a cardiac arrest and you are at an emergency room, you would want the ER doctor to have 100 percent authority to make all the decisions himself, with the others following their lead. If the doctor needed to seek consensus for what action to take amongst the nurses, administrative staff, janitors, and other hospital staff before taking action, the patients would be long dead by the time the decision was made.

However, in other circumstances, it could help with faster decision-making within an organization, when decision-making power is 'delegated at the edges.' This is often the case within Special Forces units, for instance, where those on the terrain have a significant amount of power to make decisions themselves, instead of having to have command line approval for everything, in comparison to regular military units. This could be described as a more 'holacratic' form of decentralization, rather than a consensus-based decentralization.

---

12 Jean-Michel Lafleur, "The Belgian Political Crisis of 2010-2011: An Analysis of the Causes and Consequences," West European Politics (2013): DOI: 10.1080/01402382.2013.761222.

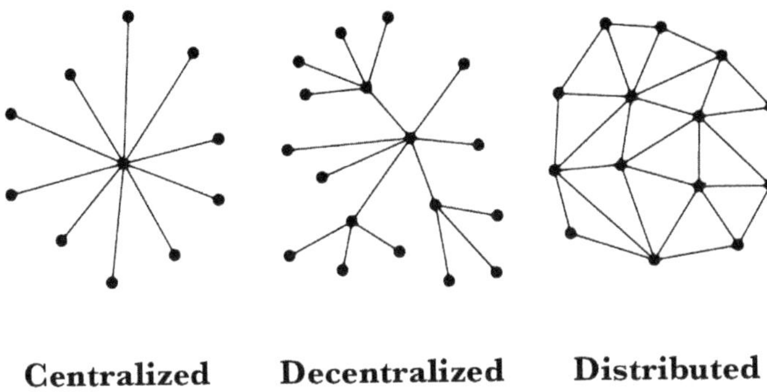

Figure 12: ***Centralized, decentralized and distributed systems.***

As we will discuss further on in this chapter, 'perfect decentralization' is rare, and it's challenging to implement at an organizational level. Hence, it's important that we use decentralization for its correct purpose: creating organizational resilience.

CASE STUDIES OF DECENTRALIZATION: REAL-WORLD EXAMPLES AND LESSONS LEARNED

Contemporary nation-state governments have gone through waves of centralization and decentralization. The word 'decentralization' was first used after the French Revolution, when the people got rid of the royal elite and became a republic. It was later used by many anarchists and libertarians in the nineteenth and twentieth centuries, and it has widely been seen as an effective way to reform governments by more mainstream segments of society as well.

In 1998, a UN report stated that: "*. . . A large number of developing and transitional countries have embarked on some form of decentralization programmes. This trend is coupled* *with a growing interest in the role of civil society and the private sector as partners to governments in seeking new ways of service delivery... Decentralization of governance and the strengthening of local governing capacity is in part also a function of broader societal trends. These include, for example, the growing distrust of government generally, the spectacular demise of some of the most centralized regimes in the world (especially the Soviet Union) and the emerging separatist demands that seem to routinely pop up in one or another part of the world. The movement toward local accountability and greater control over one's destiny is, however, not solely the result of the negative attitude towards central government. Rather, these developments, as we have already noted, are principally being driven by a strong desire for greater participation of citizens and private sector organizations in governance."*

Doctrines called the 'new public management' were published, which brought together the left and the right in the Western hemisphere, through a mixture of private contracting and increased regional participation in governance functions and decision-making. Margaret Thatcher was in the lead, as the most vocal proponent of it in three major areas: localizing, decision-making, and outsourcing. Indeed, decentralization as a term and a method that unites both the left and the right has stood the test of time, and that remains true to this day, even though, as mentioned earlier in this chapter, the term is often used in such a broad, sweeping manner that it sort of loses its actual meaning and fundamental purpose.

The term 'decentralization' gained a new sort of renaissance in the wake of the internet. When you think about it, the internet itself is very decentralized in the sense that anyone can go online and use the website of their choice or start their own. During the early years of its popularization, in the 1990s and 2000s, there was a sense of great freedom online; everyone could do pretty much what they wanted, with little to no regulation. This empowered an entire era of seeing the world through a lens of unprecedented personal liberty and global access. Over the following years, that liberty shrank gradually, as nation-states started to impose more controls over who could access what information. Most people have heard of the "Great Firewall of China,"[13] of course, tightly controlling what information China's citizens can access, banning certain search terms like "Tiananmen Square demonstration," "Hong Kong protests," "Ai Weiwei arrest" and others, and banning certain platforms altogether, such as Twitter, Facebook, and Google Search. It's a dystopian technology that Beijing has successfully exported to others, including Iran, Vietnam, Cuba, Zimbabwe, and Belarus, though some people use VPNs and other tools to get around the constraints and access the 'international internet.'

Outright public 'wars' have broken out between internet activists and various sectors at various times; take, for instance, the Pirate Party movement, founded by Rick Falkvinge, which was started in 2005 as sort of a joke political party in Sweden. Due to feuds over copyright infringements for music, movies, software, and games, the government called online pirates "pirates" on the news, meaning it as a derogatory term, which fueled self-proclaimed pirates to stand up in public with pride. What started as an offhand comment on an obscure online forum spread worldwide, with local chapters in more than seventy countries; seats in the EU, Swedish, German, Icelandic, and Czech parliaments; and several prominent members of The Pirate Bay—the file-sharing website that fueled the community—arrested and jailed. Similar wars followed with various hacktivist groups such as Anonymous, Lulsec, and Chaos Computer Club. Although online discourse around piracy has largely died down after the entrance of platforms such as Netflix and Spotify, online piracy itself is still alive and well, with millions of people sharing files on The Pirate Bay daily.

13 Lena L. Zhang, "Behind the 'Great Firewall': Decoding China's Internet Media Policies from the Inside," Volume 12, Issue 3: DOI: 10.1177/1354856506067201.

THE WEB2 AND WEB3 SHIFTING LEGACIES

The evolution of Web: A Comparative Analysis:

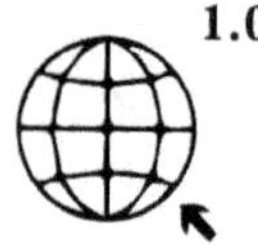

Simple banner Ads were seen

Basic Content Delivery Networks (CDN) were introduced

Personal websites were rare

Page views and screen scrapping were introduced

Contents had directories or taxonomy system

Automatic texts, graphical and interactive ads replaced simple banners

B2B and P2P file sharing protocols were invented

Bloging became popular

Cost per Click (CPC) and web services replaced older methods

Website, image and video tagging (folksonomy) replaced taxonomy system

AI personalizes Ads

Cloud based file sharing systems prevails

Semantic web will replace wikipedia

User-centric livestreams replaces blogs

AI based search engines sorts out search results

P2P user engagement replaces their ancestors

Figure 13: ***The Evolution of Web***

When Web 2.0 came along, it was hailed as the 'interactive era' because of social networks such as Facebook, Instagram, YouTube, and LinkedIn, ridesharing apps like Uber and Lyft, task delegation websites like TaskRabbit, Fiverr, and Upwork, and so on. However, eventually that too started to feel limited, as most of the traffic simply went through a few Silicon Valley giants controlling the user experience and access. Now, nearly all traffic goes through what's called 'the stacks': Amazon, Apple, Google, Facebook, and Microsoft. Of course, it's still pretty much possible to do whatever you want to do on the web, but most of us use the stacks out of comfort. What's the point of setting up your own email server when there is Gmail, for instance? Or hosting your own website when there's Amazon Web Services? For most people, there aren't a lot of upsides to doing it yourself.

Of course, there have always been alternative currents to the mainstream ones, projects intended to safeguard individual liberty and privacy online through open-source protocols and decentralization, such as the file-sharing site BitTorrent and the Linux operating system, for instance, both of which have had durability and success in the industry. However, the real mainstream 'renaissance' of this—leading up to what people call Web3 today—was the advent of the creation of the first cryptocurrency, Bitcoin.

"Governments are good at cutting off the heads of centrally controlled networks like Napster, but pure P2P networks like Gnutella and Tor seem to be holding their own."

— **Satoshi Nakamoto**, inventor of Bitcoin

Bitcoin was invented in 2008 by an anonymous developer or group of developers using the name 'Satoshi Nakamoto.' In August 2008, the domain name bitcoin.com was acquired, and in October 2008, Satoshi shared a paper titled "Bitcoin: A Peer-to-Peer Electronic Cash System" with a cryptography mailing list. The open-source code itself was released to the world in January 2009. On January 3, 2009, Satoshi mined the first block and timestamped, "The Times 03/Jan/2009 Chancellor on brink of second bailout for banks," which is interpreted as a commentary on fractional reserve banking leading up to the financial crisis of 2008/2009.[14] Initially, Bitcoin was mostly adopted by dark market operators and cryptography enthusiasts, but Bitcoin soon found another fanbase amongst libertarians who wished for less state and bank interference in their lives.

14 Benjamin Sherry, "What Is the Genesis Block in Bitcoin Terms?" Investopedia, updated November 05, 2023, https://www.investopedia.com/news/what-genesis-block-bitcoin-terms/.

From there on, it grew into many sorts of different audiences—from people who viewed the blockchain as a tool to reinvent many functions of society for a fairer and more inclusive world, to investors looking for new investment opportunities. Today, nearly everyone has heard about Bitcoin, even though adoption is still very unevenly distributed.

Bitcoin is widely hailed as the modern pinnacle of decentralization. However, it does not come without its quirks and limitations. As noted in the beginning of this chapter, there's rarely such a thing as 'perfect decentralization.'

Although the Bitcoin network is decentralized, the code base is not. It's open source and public, meaning everyone can review the code and suggest changes to it, but only a few individuals have the power to merge the suggested code changes into the main repository. Significant changes, such as halving events when the mining difficulty of Bitcoin doubles roughly every four years, require the majority of miners to agree and adopt the new version.

However, Bitcoin is holacratic in the sense that anyone can fork the code (i.e., copy it) under the MIT License, make their own version of the coin, and see if it gains traction amongst users. Some do it and simply rebrand it, while others make changes to the protocol. In that regard, it resembles more a 'swarm stigmergy philosophy,' like a flock of birds or bees flying in one direction, then one bird might take another direction spontaneously, and other birds can follow that bird if they wish to, but they don't have to.

Bitcoin unleashed a great avalanche of creativity across the world. Soon, protocols were invented for a range of other applications as well. Namecoin's aim was to make a decentralized domain name registry, file coin to provide decentralized storage, and colored coins to peg a value of any object (like a house or a car) to a cryptographic token. When Ethereum, the second-largest cryptocurrency to date, was created in 2014, it championed what it called 'smart contracts,' a term first proposed by Nick Szabo, the inventor of Bitcoin's predecessor Bitgold, which was never implemented, in 1994.[5] Smart contracts are programs stored

15 Alessandro Parenti, and Marco Billi, "Building Trust in Smart Legal Contracts," in

on the blockchain that run when the predetermined conditions are met. Smart contracts are typically used to automate the execution of an agreement so that all participants can be immediately certain of the outcome, without the need for any intermediary.

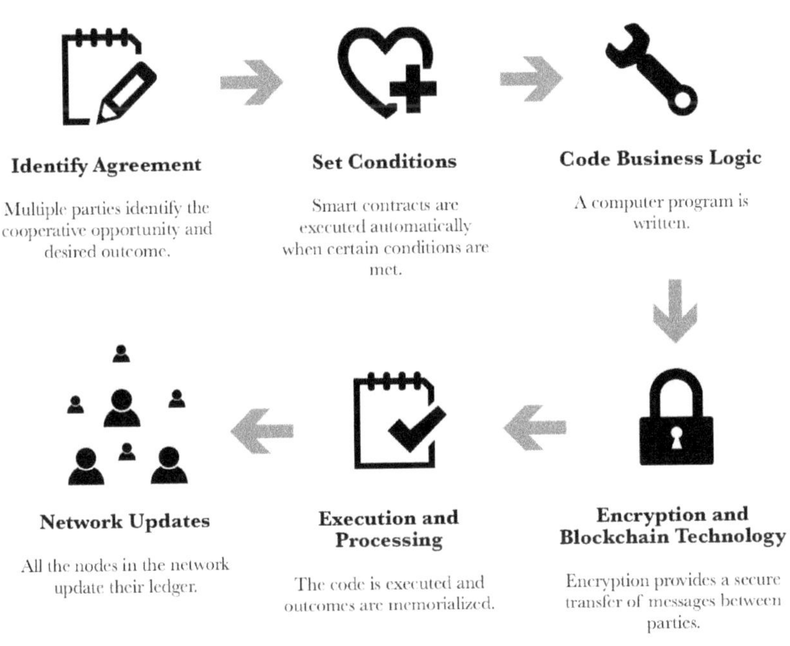

Figure 14: ***How does a smart contract work?***

Smart contracts, in turn, unleashed another creative avalanche in the form of decentralized autonomous organizations (DAOs). Other names were used for slightly different purposes as well, such as decentralized autonomous corporations (DACs) for leaderless corporations, decentralized organizations (DOs) for

Transparency or Opacity, pp. 95-114, Nomos Verlagsgesellschaft mbH & Co. KG, 2023.

organizations that were decentralized but not autonomous, and decentralised borderless voluntary nations (DBVNs) a Bitnation native term that specifically applies to DAOs designed as governance service providers (e.g., virtual nations or network-states). However, for the sake of simplicity, in this chapter, let's stick to the more widely used terminology, DAOs.

DECENTRALIZED AUTONOMOUS ORGANIZATIONS (DAOS) AND SIMILAR FORMULAS

DAOs are now common, particularly in the world of decentralized finance (DeFi). There are tens of thousands of DAOs with billions of dollars under management. It clearly seems and feels like the future, but technology and culture are still in their early infancy, leading to countless challenges. Some of the challenges lie in the decision-making processes, where token holders vote on protocol changes. One primary issue is voter collusion; in some cases, larger token holders can have more votes. Some DAOs are solving that through ID checks or NFT-based memberships, in an effort to attempt a one-person, one-vote system. Another even greater challenge for DAOs currently is voter apathy. DAOs, in general, have record-low voter participation. Even DAOs that try to solve voter apathy through a system of delegates, where people can delegate their vote to a representative whose opinions they sympathize with, are equally plagued by voter apathy because delegates change often and are usually not very active either.

A few examples of notable, non-DeFi DAOs are:

Ukraine DAO: Ukraine DAO was established in February 2022 by the Moscow-based punk band Pussy Riot to help fund the Ukrainian defense against the Russian invasion. The DAO has raised over US$7 million in crypto, which it has donated to the Ukrainian government, armed forces, and other organizations.

VitaDAO: VitaDAO is a longevity research DAO that gives out research grants to 'death valley' research in bioscience (i.e., projects at an early idea stage with high failure rates). They've raised US$4.1 million to date, including from Pfizer.

Gnosis: Gnosis started as a decentralized prediction market, where participants can vote on what they think is more likely to occur—in sports, politics, or marketplaces. It has since evolved into an infrastructure provider for the entire Ethereum ecosystem, notably creating one of the most used multi-signature wallets, the open source and free-to-use Gnosis Wallet.

These are just a few examples from a long list of DAOs that aim to fill various governance and public goods-related fields, and their number and impact are growing each year.

In general terms, even though the knowledge and ownership of cryptocurrencies is widespread today, day-to-day usage is very low. Few people and enterprises accept cryptocurrencies as payments because they are still perceived as volatile and complicated. Exchanges have been riddled by scandals and hacks, and wallet seed-phrase self-custody is hard even for technologically enthusiastic persons. Even though the technology exists, it will have to become safer and more user-friendly for widespread day-to-day life adoption. This is crucial to work toward because cryptocurrencies and blockchain technologies are fundamental building blocks for a better, fairer and more inclusive society.

DECENTRALIZATION AND BITNATION: HOW DECENTRALIZATION FORMS THE CORE OF THE BITNATION MODEL

The idea of Bitnation in itself—the idea of a system of limitless governance service providers, that everyone should be able to create their own nation, and should be free to join and leave nations at will—is an attempt to radically decentralize the entire governance system worldwide.

To that end, Bitnation was designed to be decentralized both in terms of technology as well as in its human structure. In the first whitepaper published in 2014, a system of 'embassies' and 'ambassadors' was outlined and then implemented. Anyone could sign up to become an ambassador and register their home or their business as an embassy. There were over two hundred ambassadors and fifty embassies around the world, some of which were voluntarily used to host refugees. Likewise, for the first years of operation, nearly everyone was a volunteer. Anyone could join the chat channel and start contributing code or marketing, organizing meetups, or working on their own governance project, and hundreds of people did so.

Bitnation was not really a startup; it was more like a political movement—or rather, a post-political movement attracting people of all ages and nationalities, from all sides of the political spectrum. That changed dramatically after Bitnation formally incorporated in 2017/2018 in order to raise funds. Bitnation had avoided formal incorporation up until then because a nation-state incorporation was perceived as a potential central point of failure to an organization. After all, it meant bank accounts that could be seized, executives who could go to trial, and so on. But it became impossible over time to work without proper funding, and most investors required a formal incorporation. In the end, Bitnation had a whopping five different corporate incorporations in five different jurisdictions, because the laws around token sales changed almost weekly, being all right in a jurisdiction one day, then banned the next. It cost a lot of money and was extremely time-consuming. But for the first time, Bitnation was able to pay employees and financially incentivize ambassadors. The salaries changed the entire internal dynamic for the worst, unfortunately. People worked less hard, were less motivated, and the movement lost a lot of its sincerity. In a way, it's fair to say that as it became a formal startup, the movement nearly died out entirely, or at least went into a very deep winter sleep. That's not to say that funding isn't a good thing—quite often, it's an essential part—but the community shouldn't be built around speculations on financial gains, which can sometimes be hard to avoid in the world of crypto projects.

But simply saying, "It's a decentralized technology and movement," isn't enough of course, and true decentralization is extremely hard to obtain. The only currencies to date that might be able to operate independently from their core are probably Bitcoin, Monero, and, by now, Ethereum as well. Because even though the code is decentralized, the infrastructure is not. Someone has to maintain the domain name, the website, the communication platforms . . . It's very well to say that a project is community-driven, but that's generally only true when things go well for the project, and the community prides itself on its success. But as soon as there is a problem, then that's the fault of the founder, and the founder alone . . . This puts a huge mental strain on founders, which is why more and more founders in the crypto space choose to either be entirely anonymous, or be public but not engage in day-to-day interactions in community chat channels. While this protects founders mentally and physically, it decreases the trust and accountability of projects, and makes them more centralized, as the human project core is separated from the vast majority of the community.

There is a natural life cycle for most online communities, such as chat groups, subreddits, forums, and so on. It's generally established that when a community starts online, it's a few passionate people engaging and adding a lot of value to the conversation, but then as it grows in popularity, it becomes more anonymous and diluted, and it tends to fill up with spam, distracting conversations, and trolling, until it gets taken over by these elements. By then, forums generally die, or they go into a strict 'community control mode,' where human and bot moderators enforce strict moderation of members and comments. Community moderation can work and is generally a must, but it's hard to get the level right without losing the sense of freedom, openness, and creativity that makes online communities so valuable to start with. This mirrors offline societal trends like Dunbar's number as well—the number of stable relationships people are cognitively able to maintain at once. The number the Oxford evolutionary psychiatrist Robin Dunbar proposed was 150 persons, and in the experience of the Bitnation organization, that's generally the number of people who were constructively active on the chat channel at any one time.[67] However, it's perfectly possible that the human brain is going through a rewiring process in that regard, as we're used to seeing so many impressions from so many people online at all times. This is why Susanne believes in the concept of holacracy: when any one group gets too large, instead of having to play internal politics through voting or signaling, sub-groups can simply form spontaneously to better suit their culture and expertise, allowing for more fluidity and intimacy. It's much like the way everyone can create their own subreddit or Facebook Group, or build their own decentralized application (DApp) on blockchains like Ethereum.

It is, of course, the hope of many startup founders, inspired by Satoshi and others, that projects will take a life of their own, independently of them, to become immortal in a way, and that hope tends to rely mainly on the promise and premise of the technological decentralization itself. As discussed above, the Web3 movement aims to decentralize everything online—not just money, but also ownership, identity, domain names, communication channels, social media channels, and much more. Bitnation went very far in that aspect, trying to decentralize

16 P. Lindenfors, A. Wartel, and J. Lind, "'Dunbar's Number' Deconstructed," Biology Letters 17 (2021): 20210158, https://doi.org/10.1098/rsbl.2021.015.

17 Robin Dunbar, How Many Friends Does One Person Need?: Dunbar's Number and Other Evolutionary Quirks (Faber & Faber, 2010), ISBN: 978-0571253432.

the entire governing order of the world, including the law enforcement system (through a decentralized AI-powered reputation bot named Lucy), law creation and arbitration, identity, and nationality. This remains Bitnation's aim today.

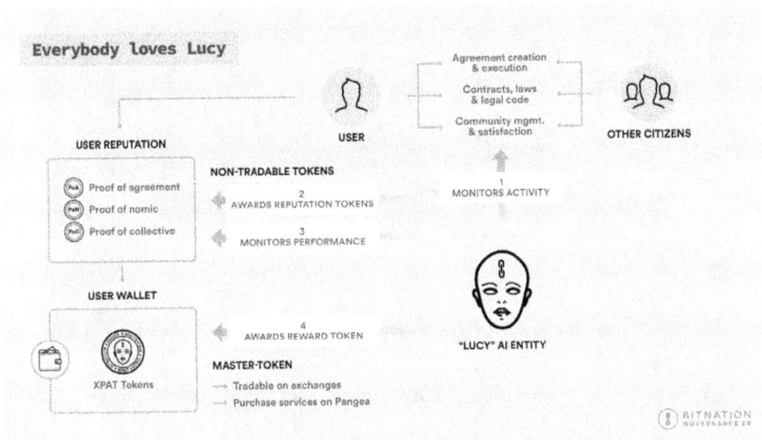

Figure 15: ***The Lucy Reputation System from a Bitnation presentation 2018***

CHAPTER THREE

TRANSCENDING BOUNDARIES: THE LEGAL CODE IN THE ALGORITHMIC MIRROR

THE ALGORITHMIC ALCHEMIST: AI'S DANCE WITH LEGAL CONSTRUCTS

The emergence of artificial intelligence has not only revolutionized industries and daily life but also deeply impacted the legal realm. Its transformative nature presents opportunities and challenges, as it is reshaping the way we perceive legal constructs and principles.

AI's intrinsic capability of parsing vast datasets and drawing patterns enables it to facilitate legal processes in unprecedented ways. For instance, AI-driven legal research tools can swiftly analyze thousands of precedents, extracting relevant legal points and making case law more accessible to attorneys. Additionally, AI-enhanced contract management tools streamline the drafting, negotiation, and execution of contracts, enhancing efficiency and reducing errors.

Yet, AI's role in the legal sphere isn't limited to assistance; it also questions fundamental legal tenets. For example, the rise of autonomous vehicles necessitates a reexamination of liability constructs. If a self-driving car causes an accident, who's to blame? The car's owner, its manufacturer, the software developer, or the AI itself? Traditional legal notions of responsibility are being reimagined in the AI age.

The interaction between AI and law also shines a light on biases and fairness. Machine learning models, if trained on biased data, can perpetuate and even amplify societal prejudices. Legal professionals and policymakers are grappling with the question of how we can ensure that AI-driven decisions, especially in areas like criminal sentencing or loan approvals, are free from undue bias and that they uphold the principle of equal protection under the law.

Further, AI's deep integration into society presents new challenges related to privacy and surveillance. With AI systems capable of facial recognition, predictive analytics, and behavior profiling, the balance between security and personal freedom is being tested. As AI continues its dance with the law, it's crucial for legal frameworks to evolve, ensuring that technological advances respect human rights and uphold justice.

Lastly, the very essence of AI—its capability to learn, decide, and act—poses philosophical questions about legal personhood. Can, or should, an AI entity possess rights or obligations? If an AI creates a piece of art or invents a new gadget, who owns the intellectual property?

In previous chapters, we established that law is a technology. Another way of putting this is to say law is a tool. In the technical sense of the word, rules can be thought of as laws. In the interest of AI safety, the human owners of this technology need a way to set rules in place that can logically determine all possible behaviors that can unravel from their AI.

The first notion we want to emphasize is the concept of correctness in software. In the instance where the AI generates the software, such as in cases of software synthesis, which was discussed for our hypothetical AI companion software,

Lucy, we would require the owner to set constraints. These constraints set by the owner guarantee AI safety by imposing rules in the case of AI, which, in theory, can be in alignment with human laws.

Throughout history, science-fiction authors have shaped history by exploring ideas that titillate engineers and entrepreneurs alike, who then fancy working on those ideas. Isaac Asimov (1920–1992) was a prolific science-fiction writer who had a lasting impact on technologists.[1] His influence was similar to the impact of writer Ayn Rand, the libertarian author who famously sold more copies of her book Atlas Shrugged in the US than the Bible sold. Isaac Asimov wrote several series of novels, novellas, and short stories; among them are notables such as the Foundation series and the iRobot series. In fact, the concept of robotics itself was invented by Isaac Asimov because he coined the word robotics.

One major contribution is Isaac Asimov's three laws of robotics. The three laws are explained:

1. A robot may not injure a human being, or, through inaction, allow a human being to come to harm.
2. A robot must obey orders given to it by human beings, except where such orders would conflict with the First Law.
3. A robot must protect its own existence as long as such protection does not conflict with the First or Second Law.

As you can see, these laws are written in such a way that they can be logically defined. These rules would essentially serve as laws for robots. In our case, if the owner of an AI companion so decides, they could set in place these laws, as well as additional laws, in order to maintain complete dominion and control. In fact, human laws or regulations could require that humans implement certain standard rules to govern their AI agents for safety reasons. In this way, the rules

18 Isaac Asimov, I, Robot (Gnome Press, 1950), ISBN: 978-0553294385.

of humanity, which govern humans, would also influence the behavior of all AI that interacts within society.

The second notion we want to emphasize is the notion of self-sovereignty. We can express this concept by saying that the person who knows the secret words enabling them to make transactions is the property owner, and in the philosophical sense, it pertains to their mind. If it is a system where multiple signatures are required to complete a transaction, then the group is in control, and as a result, it is the collective minds of that group acting in unison.

SELF-SOVEREIGNTY IN THE DIGITAL ERA

In the age of smartphones, smart contracts, and private keys, self-sovereignty is being reimagined. Today, personal autonomy is deeply intertwined with the digital domain, reshaping how we conceptualize identity, privacy, and freedom. In the digital realm, identity has evolved from tangible, physical tokens to intangible, knowledge-based proofs. Your smartphone, often secured by a personal code or biometric data, becomes an extension of yourself. It contains your photos, messages, apps, and more, representing a digital mirror of your preferences, communications, and experiences. Similarly, private keys and passwords, whether they are alphanumeric strings or a set of twelve words, serve as unique identifiers. They grant access to your digital assets, online profiles, and more. The exclusivity of this knowledge—the fact that only you should know your passwords or private keys—makes it a defining attribute of your digital identity. Just as your physical presence is verified by unique, innate characteristics like fingerprints or retinal patterns, in the digital space, you are recognized and authenticated by the exclusive secret knowledge you possess. This intangible proof solidifies your presence, ownership, and actions online, making "you" an amalgamation of your physical self and the unique digital keys you hold.

One of the key contributors to these ideas is open-source software advocate and technology writer Doc Searls. He is a co-author of *The Cluetrain Manifesto* and is the author of *The Intention Economy: When Customers Take Charge.*[23] Doc Searls is a longtime advocate for customer empowerment. It was Doc Searls who introduced the concept of intent casting, which means formulating your intent in a way that a machine can understand it. Similar to how people use prompt 'engineering' for AI software, such as text-to-text prompts (ChatGPT), text-to-image prompts (Midjourney, Firefly, etc.) or text-to-video prompts. With intent casting, or 'prompt engineering', it puts the outcome of the software, the formation of the software back into the hands of the end user.

Doc Searls' vendor relationship management (VRM) concept marks a pivotal shift in the dynamic between consumers and service or product providers. VRM seeks to invert the traditional customer relationship management (CRM) paradigm. Instead of businesses managing their relationships with customers, VRM tools allow consumers to manage their relationships with businesses. This means giving individuals control over the data they share, who they share it with, and under what terms. This concept is implementable and may be a default in Web3 due to it being decentralized, but in Web2, the norm is to centralize data into silos owned by vendors.

In the context of the aforementioned discussions on AI, legal systems, and the extended digital self, VRM plays a vital role in data sovereignty and personal agency. As individuals' lives become more intertwined with digital spaces, the data generated becomes a critical component of their identity and autonomy. AI in legal systems, for instance, relies heavily on data for its functionalities. If individuals employ VRM tools, they can control and decide which data an AI system can access, ensuring that personal data isn't misused or accessed without clear consent. Moreover, VRM's empowerment of individuals to dictate terms can

19 Rick Levine, Christopher Locke, Doc Searls, and David Weinberger, The Cluetrain Manifesto: The End of Business as Usual (Basic Books, 2001), ISBN: 978-0465024094.

20 Doc Searls, The Intention Economy: When Customers Take Charge (Harvard Business Review Press, 2012), ISBN: 978-1422158524.

also influence how AI systems are designed, ensuring they respect individual data rights and prioritize user-centric approaches. Thus, VRM stands at the intersection of individual digital rights, AI's data-driven functionalities, and the evolving landscape of digital interactions, acting as a linchpin that can help balance power and ensure user agency in an increasingly digital world.

PRIVATE KEYS: KNOWLEDGE AS OWNERSHIP:

$$h = \text{hash}(\,data\,)$$

At the heart of this transformation is the concept of the private key. A private key is a unique string of numbers and letters that serves as both an individual's identity and the key to their digital vaults. Unlike traditional forms of ID or asset possession, which are physically held or stored, a private key is pure knowledge. If Alice or Bob knows a private key, they own whatever assets or identities it unlocks, whether that's cryptocurrency in a digital wallet or access to a digital identity. If they forget it, or it's stolen without recovery, they lose access. The entire notion of ownership in this realm is predicated on the exclusivity of this knowledge.

The knowledge, which are your seed words representing your private key, can be lost. Currently, there is a situation where knowers must make trade-offs between convenience and security. There are methods that can mitigate this, such as the use of Shamir Secret Sharing (SSS), in combination with encrypted cloud storage. Additionally, there is a secure enclave that securely holds the information in encrypted form on a hardware device, such as a smartphone or hardware wallet.

Shamir's Secret Sharing is a cryptographic algorithm created by Adi Shamir, which allows a secret to be divided into parts, giving each participant their own unique part.[4] To recover the original secret, a minimum number of parts (thresh-

21 Dan Bogdanov, "Foundations and properties of Shamir's secret sharing scheme research seminar in cryptography," University of Tartu, Institute of Computer Science 1 (2007).

old) is required. This is commonly referred to as a (t,n)-threshold scheme, where t is the threshold number of parts required to reconstruct the secret, and n is the total number of parts or shares that the secret has been divided into.

Ledger Wallet, a hardware "wallet" company, is the first that we are aware of to get the balance right.[5] They offer a feature called Ledger recovery, and the question comes in how this feature is implemented. Shamir Secret Sharing is secure, allowing the private key to be broken into shares. The next step would be to select for redundancy and distribute the shares to multiple cloud services. If these shares are double encrypted, you can put them anywhere.

Additionally, multisignature "wallets" allow for security with recovery. If you have threshold signatures, where you need to sign from multiple devices to transact, this is secure and convenient for many use cases. The only issue is that the average person does not have multisig setup, does not follow best practices, and does not have a hardware "wallet". Therefore, for people like this, Ledger recover might be the only option, and for those who do not have a ledger, Coinbase [1] [2], Binance, or a traditional bank may be what they use. It must be noted that if crypto is stored on these platforms, you do not know the private key, and your crypto can be confiscated by whoever does know the key.

TRANSACTIONS AS FREE SPEECH:

$$\text{Value}_{\text{final}} = \text{Value}_{\text{initial}} - \text{Amount}_{\text{transferred}}$$

Every transaction made using smart contracts or digital currencies is essentially a declaration—a form of expression. When Bob sends Alice a cryptocurrency, he's broadcasting a message to a network, saying he transfers value to Alice. It's the modern-day digital equivalent of announcing a declaration in a

22 https://www.ledger.com/.

public square. This act of broadcasting or transacting is a form of free speech. It's an assertion of one's intent and autonomy without the need for intermediaries, on a blockchain that is intended to be permanent, immutable (extremely difficult to modify retroactively) and public.

SMARTPHONES – EXTENSIONS OF THE MIND:

Cognitive Reach = Human Capacity + Smartphone Functionality

Smartphones have revolutionized our connection to the digital world. They aren't just communication devices; they are extensions of our minds. Just as a prosthetic limb might extend the physical capabilities of its user, a smartphone extends our cognitive and communicative reach. With it, we can recall facts, communicate across vast distances, and even establish our digital self-sovereignty by holding our private keys.

People often speak about brain-to-tech interfaces, meaning things such as chips implanted in the brain and so forth, but at present, smartphones are a de-facto brain-to-tech interface, and a very tactile and powerful one, to boot. It will be interesting to explore where the venture between Meta, the company behind Facebook, and RayBan, the eyewear fashion brand, goes in its quest to create 'smart glasses.' Perhaps Apple's Vision Pro will have more success than its predecessor, Google Glass, which never made it past beta testing.

AI AGENTS: EXTENDING SPEECH:

Communication Amplification = Human Output
x AI Enhacement Factor

Artificial intelligence, especially chatbots and digital assistants, have begun to act as extensions of our speech and expression. When we task an AI to communicate on our behalf, whether it's setting up appointments, responding to emails, or engaging in digital negotiations, we're extending our voice through technology. In essence, an AI agent amplifies, refines, and represents our intent, much like how a megaphone amplifies our voice in the physical world.

Some people might argue that technology removes us from each other, but the idea that it amplifies us is a much more accurate summation. If we venture back to the early days of the internet, to the era of Web 2.0 and social networks, amplifying ourselves was exactly what we did. For the first time, many people felt they could reach out and communicate in their true fashion beyond a locally restricted world that didn't necessarily understand them. They saw that elsewhere, there were people who might understand what they believed or who might accept them as their true selves.

As mentioned in the first chapter, David Chalmers and Andy Clark introduced the extended mind thesis, proposing that our minds aren't confined to our brains but extend into the external world through tools, artifacts, and even interactions with other people. When individuals use external elements such as digital tools, their cognitive processes can be viewed as 'extending' into those mediums. Building on this, when multiple extended minds collaborate, they form interconnected 'mind networks,' where shared cognition takes place across various tools and individuals.

Artificial intelligence can amplify these mind networks. This is especially true for logic-based systems like Cyc, a product from Cycorp, a company known for using symbolic AI. Cyc, designed to emulate human common-sense reasoning, can integrate diverse expertise from multiple minds, facilitate communication between them, and dynamically adapt to new information.[6] Essentially, AI serves as a bridge, enhancing the interconnectivity of extended minds and fostering a more profound collective intelligence.

23 Cycorp, https://cyc.com/.

Building on the concept of identity in the digital realm, the idea of the extended mind posits that our cognitive processes and sense of self are not solely contained within our biological brains but can extend into our tools and technologies. When we assimilate digital devices like smartphones into our daily lives, or when our identity and assets are tied to intangible proofs like private keys, our mind essentially integrates with these digital extensions. This seamless merger of human cognition with technological apparatus reshapes our identity and capacities. In essence, this integration transforms us into cyborgs: beings where the organic and the synthetic coalesce, extending our capabilities and redefining our very nature in the interconnected digital age.

Artificial intelligence offers transformative potential for the legal system. It can expedite legal research by analyzing vast archives of documents and cases, predicting possible case outcomes based on past rulings, and streamlining processes such as document reviews, contract analysis, and even basic legal assistance through chatbots. AI-driven tools can also enhance administrative efficiency within court systems, aid in bias detection by analyzing historical decisions, and provide valuable analytics for law firms aiming to improve their service delivery.

THE REALM OF MANY CENTERS: POLYCENTRIC LAW IN THE DIGITAL AGE

In traditional legal structures, we are accustomed to a singular, centralized authority, be it a national government or an intergovernmental organization, that creates, enforces, and interprets the law. However, as the digital age progresses, a more fragmented yet interconnected paradigm is emerging polycentric law. This framework, characterized by multiple overlapping jurisdictions and centers of governance, challenges conventional notions of legal territoriality and sovereignty.

The term 'polycentric' itself means 'having many centers,' suggesting a landscape where various legal entities or communities coexist and operate concurrently. In the digital realm, this takes shape in the form of online communities, decentralized autonomous organizations (DAOs), and initiatives like Bitnation that craft their own codes of conduct and mechanisms for conflict resolution.

Polycentric Law: A Landscapte of Legal Diversity

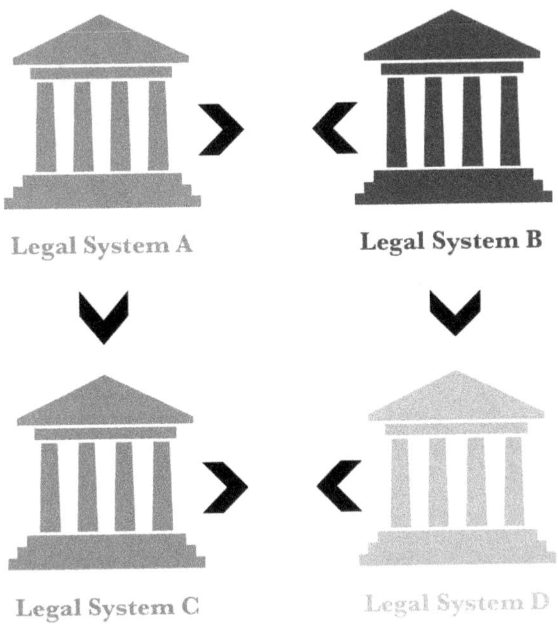

Figure 16: ***Multiple legal systems coexist, offering choice and adaptability.***

Bitnation, a decentralized borderless "voluntary nation", exemplifies this polycentric approach. It empowers individuals to design and choose their own personalized legal systems and governance structures. Leveraging blockchain technology, Bitnation offers services like contract creation, dispute resolution, and even marriage registration, independent of traditional nation-state constraints. Participants align with the systems they find most appealing, making governance a bottom-up rather than top-down process.

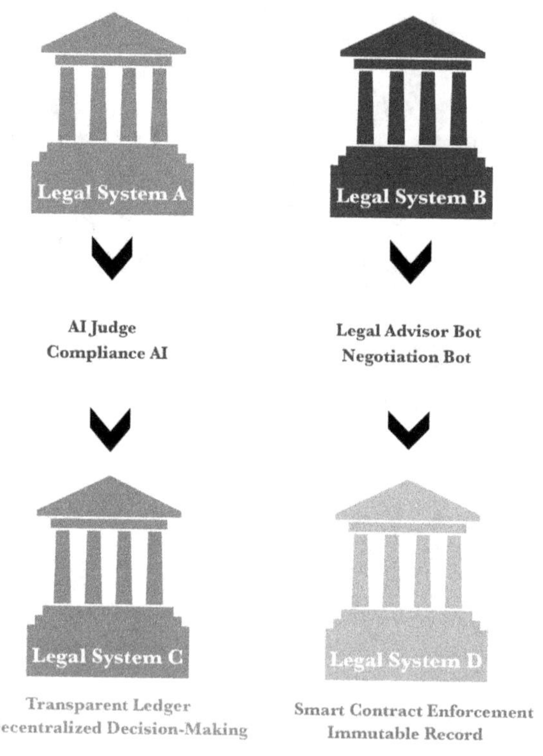

Figure 17: ***Integrating advanced technology into the diverse legal landscape***

However, the rise of polycentric law presents as many challenges as it does opportunities. One glaring concern is the potential for legal confusion and conflict. With myriad jurisdictions and rule sets, whose law takes precedence when disputes arise between members of different digital and geographical communities? One would imagine that contract law will be rather straightforward, as it's in all parties' interest to solve the conflict; however, criminal law will be far more challenging when it comes to agreeing to a jurisdiction, for instance, if someone has been robbed or raped. Moreover, it remains unclear whether decentralized entities like Bitnation can ensure consistent legal protection and uphold human rights standards.

Despite these challenges, the allure of polycentric law lies in its inherent flexibility and adaptability. As opposed to rigid, one-size-fits-all legal systems, polycentric structures can evolve rapidly, adjusting to the diverse needs and values of their members. This dynamic nature can lead to innovative solutions to legal challenges, tailored to the digital age's unique demands.

Furthermore, polycentric law provides an avenue for marginalized groups or those disillusioned by their nation-states' legal systems to find or create communities more aligned with their values and aspirations. It's a form of 'jurisdictional competition,' where different legal systems vie for participants based on their efficiency, fairness, and relevance.

For polycentric law to be genuinely effective in the digital age, a degree of interoperability is essential. Different legal systems must find ways to communicate, collaborate, and coexist, ensuring that the digital realm remains cohesive amidst its diversity.

In sum, the emergence of polycentric law signals a profound shift in the way we conceive governance and legality in the digital era. As platforms like Bitnation continue to push the boundaries of what's possible, we're prompted to reimagine the very fabric of legal systems, embracing a future that's as multifaceted and interconnected as the digital world itself.

AN UNFOLDING SAGA: LARGE LANGUAGE MODELS AS SCRIBES OF THE LEGAL PROSE AND THE SYMPHONY WITH SYMBOLIC AI

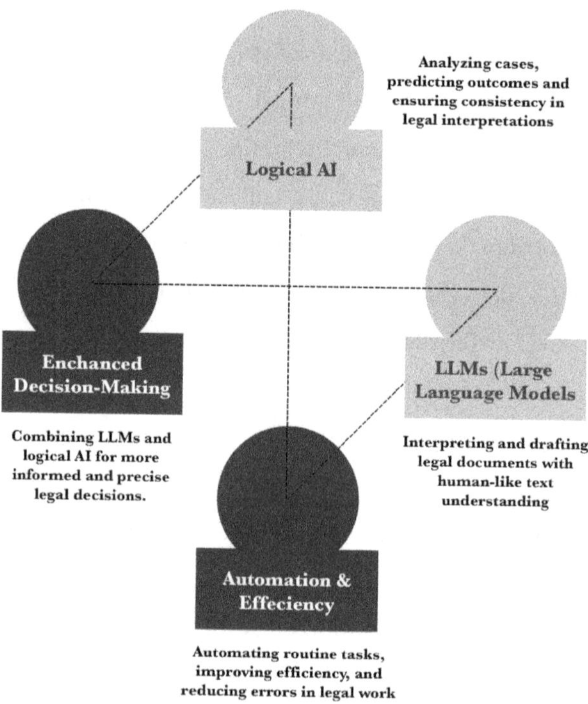

Figure 18: ***Revolutionizing legal processes through advanced technologies.***

The emergence of large language models (LLMs) has painted a vivid picture of the future of law. But to understand the complete range of possibilities, one must also turn their gaze to another realm of AI: symbolic AI, exemplified by systems like Cyc, a product created by Cycorp. Where large language models thrive on vast quantities of text, symbolic AI operates on explicit representations of facts, rules, and logic. The former excels in understanding and generation, while the latter boasts deductive prowess and reasoned decision-making.

The potential advantages of symbolic AI in the legal domain are manifold. Given that law often operates on a foundation of codified rules and principles, symbolic AI can directly model these rules and apply them consistently. For instance, determining the applicability of a particular law to a given case can be effectively managed by symbolic AI, tracing through predefined logic trees and delivering reasoned conclusions.

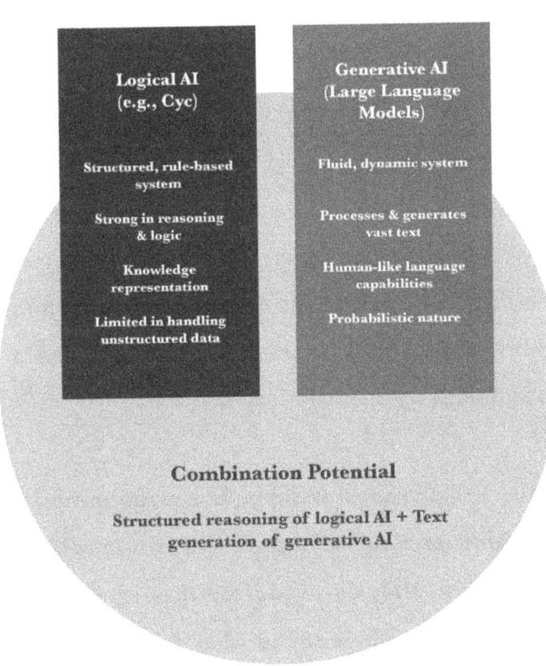

Figure 19: ***Combining Logical AI with Generative AI***

Now, imagine the magic that can be conjured when the strengths of both these AI paradigms are harmoniously combined in the legal arena. Large language models could trawl through the expansive legal literature, synthesizing information, and then feed relevant insights to symbolic AI systems. These systems, in turn, could apply their rule-based logic to deduce outcomes or recommend actions. The resulting collaboration would blend the best of both worlds: the intuitive grasp of language models with the structured reasoning of symbolic AI.

This confluence could pave the way for AI-assisted legal platforms with unparalleled capabilities. For instance, while drafting a contract, a language model could generate the necessary clauses, and a symbolic AI could then check their validity and consistency with existing legal frameworks. In litigation scenarios, language models could predict potential arguments based on historical data, while symbolic AI evaluates the strength of each argument based on codified legal principles.

However, as with any innovation, challenges persist. The integration of two distinct AI methodologies would require sophisticated architectures and rigorous checks to ensure that the systems' outputs align with legal and ethical standards. Moreover, the perennial concern of transparency becomes even more pronounced; understanding the decision-making pathway in such a combined system might be complex, necessitating tools and interfaces that elucidate this process for users.

While the promise of large language models in law is immense, the true frontier may lie in their symbiotic relationship with symbolic AI. Together, they can redefine the legal landscape, offering tools that are not only efficient but also deeply insightful, reasoned, and just. The intersection of these AI paradigms allows for a legal future that's not just automated, but also profoundly intelligent.

CHAPTER FOUR

THE EVOLUTION OF GOVERNANCE AND THE WESTPHALIAN APARTHEID

GOVERNANCE SYSTEMS—OUR MEANS OF ORCHESTRATING SOCIETAL ORDER AND PROGRESS—ARE A MULTIFACETED TOOLBOX.

They encompass everything from basic structures that enforce laws to intricate mechanisms that safeguard the rights of individuals. It is the complex interplay of rules, norms, institutions, and power dynamics that defines the boundaries within which a society functions. Whether it is a sovereign state exercising authority over its citizens or an emergent decentralized network-state, the fundamental principle remains: governance is the foundation upon which the edifice of society rests.

In the grand tapestry of human history, the concept of governance has woven itself as an intricate thread, shaping the destiny of societies across the ages. From the primitive tribes to the many sprawling nations of today, the evolution of governance is a story of adaptation, negotiations, and the paradoxes that emerge from it. It can feel as if our current nation-state system has emerged organically throughout time in a sort of linear fashion, and that it is a relatively well-functioning end state, but nothing could be further from the truth. In fact, the nation-state system as we currently know it is only about four hundred years old, a short blip in human history.

HOW THE WESTPHALIAN MODEL CREATED A STAGNANT GLOBAL APARTHEID

To understand the evolution of governance, we must journey back to our ancient ancestors. The dawn of governance can be traced back to the very moment when humanity realized the advantages of cooperation. In early communal societies, decisions were likely made through informal consensus, guided by shared norms and needs. With the advent of agriculture, the stakes grew higher. The need to manage resources, resolve disputes, and ensure survival prompted the emergence of leaders and councils to arbitrate and legislate. This marked the inception of a more formal form of governance, a response to the complexities of organizing growing populations. As societies grew, governance took different forms, from tribal councils to feudal systems—often crystallizing around charismatic leaders or religious figures. The rise of centralized empires like the Roman and Byzantine empires brought a shift towards more encompassing governance structures.

The Westphalian Peace of 1648 marked a turning point in governance. By recognizing the sovereignty of nation-states and promoting non-interference in domestic affairs, it laid the groundwork for modern international relations.[1] This

24 Andreas Osiander, The Westphalian System (London: Routledge, 2001), ISBN: 978-0415258813.

system centralized authority within defined borders, providing a framework for diplomatic relations and conflicts. Yet, as we shall explore, this very centralization sowed the seeds of the Westphalian paradox—the current nation-state system we live in.

The Peace of Westphalia, which emerged from the devastation of the Thirty Years' War, sought to bring stability to Europe by establishing a system of sovereign states. Each state was recognized as the ultimate authority within its territory, a principle that aimed to prevent further religious and territorial conflicts. This shift from the dominance of the Catholic Church to the sovereignty of secular rulers marked a significant change in governance philosophy. The Westphalian system, while contributing to a reduction in religious conflicts, also inadvertently reinforced the notion of fixed borders and exclusive sovereignty. A 'nation-state' is a distinct concept from both a state and a nation. The official definition of a nation is a large type of social organization where a collective identity has emerged from a combination of shared features across a given population, such as language, history, ethnicity, culture, territory, or society. A state, on the other hand, is a governing body that administers a specific geographic area. This explains why the name Bitnation was chosen, rather than Bitstate, as a nation does not imply territory. This also makes the definition of 'network-states' a little confusing.

Hence, what the Peace of Westphalia eventually led to over time was the roughly 195 Nation-states we have today (it's slightly more or less, depending on whose measurement you use to count them). These 195 nation-states now cover all of Earth, except for Antarctica, the high seas, and a few tiny disputed areas on land. That effectively means that there's no way to opt out of the current nation-state system if one does not consent to it.

The 'voting with your feet' analogy does not work for many reasons. First of all, the vast majority of people in the world do not have a passport that allows them to easily move to another country, and that's not even taking into account the estimated eleven million stateless people in the world. Second, for many, it's not doable for other reasons as well: language barriers, family obligations, health conditions, economic situation, and other obstacles are common. And many might simply not want to leave because they're attached to their local culture or social circle. But let's say even if you could technically leave your place of birth and choose whichever country you wanted to, there are still only 195 countries to choose from for a population of approximately eight billion people. Imagine if there were only 195 websites in the world, or 195 recipes to cook, or 195 movies to watch. Why do we accept such limited options for something so important in our lives?

Nation-states do, in general, not compete for their citizens, since it's difficult for people to leave, and nation-state governments will be paid for their services whether people want them or not, through extracting involuntary taxes. Failure to pay for services, regardless of whether you have voluntarily agreed to them or not, will, for the most part, land you in jail or with huge fines. Hence, unlike businesses and other private-sector actors, nation-states don't have any need to provide better services at better prices in order to survive as an organization in a competitive marketplace. For instance, a company producing laptops or cell-phones needs to constantly innovate in product quality or in price value proposition to stay ahead of its competitors, which has led to the rapid development of tech products. Governments should innovate in the same way, and for the same reason.

People often quote the democratic process as being the most effective tool to keep politicians in check and make them compete against each other. However, in most nation-states, there are only elections every four or five years, there are generally very few candidates to choose from, and often it only changes the top layer of the government. On top of that, only one side wins, meaning everyone has to conform to a single solution. Further, it leads to populist leadership, in which leaders care more about themselves or their party being re-elected in the next cycle than about working on resolving hard long-term societal issues.

Like the late American political activist **R. Lee Wrights** eloquently said:

"Democracy? I want nothing to do with a system which operates on the premise that my rights don't exist simply because I am outnumbered."

For those concerned with human rights, the Westphalian system should be of even greater concern. Most people agree that people's opportunities and quality of life should not be determined by things they cannot choose, such as gender, skin color, or sexual orientation. Yet, we are living in a gigantic apartheid, where the geographical area where you were arbitrarily born entirely defines your life. Those with 'good passports' have opportunities, while those with 'bad passports' have far fewer opportunities and might even be trapped in a situation of extreme poverty, warfare, and premature death. It does not have to be like that.

GOVERNMENTS, CORPORATIONS, NON-GOVERNMENTAL ORGANIZATIONS, OTHER FORMS OF GOVERNANCE

A government is a set of people who do a set of actions according to a set of interests and principles. It is not, in any way, a magical entity—there is nothing that a government can do that another group of people cannot do. The only difference between a government and most other groups in society is that they use force, or the threat of force, to carry out their work. Any other group could provide the same services, such as a private company, local community network, or non-profit organization, and they would do it more efficiently as well since they need to compete for citizens, or customers, to guarantee revenue for themselves. And they do; many governance services are not provided by governments, and in many ways, governments provide services that are generally not associated with governance, but rather with the private sector.

Perhaps one of the most blatant examples of this is government bonds. Like an exchange listed company, nation-state governments do indeed sell shares of projected future profits to whoever cares to invest in them. In some rare instances, governments have even sold war bonds, for a return on wars carried out by the nation. The US government was famous for doing this during the world wars, and most recently, Ukraine, has been doing it to finance its defense against Russia. Governments credit ratings can be upgraded or downgraded based on how the market perceives their economic viability, which often has a significant impact on the average citizen, through inflation and mutual pension funds placed in bonds.

Figure 20: ***Replica of a U.S. War Bond Propaganda poster from WWII***

Furthermore, depending on the country, many governance services are not carried out by governments. More often than not, private contracting companies build roads, weapons, and infrastructure. In North America and Western Europe, many government-financed healthcare and education providers are private institutions and practitioners.

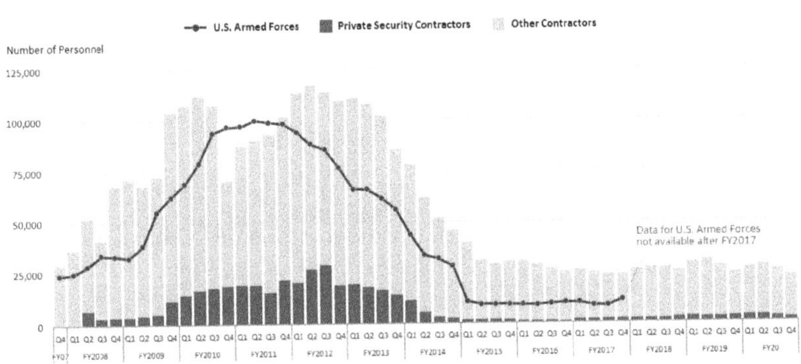

Figure 21: **U.S. Armed Forces and DOD Funded Contractor Personnel in Afghanistan Q4 FY2007-Q4 FY2020** *Source: Created by CRS. Contractor levels drawn from USCENTCOM Quarterly Contractor Census Reports; U.S. Armed Forces levels through Q4 FY2017 drawn from "Boots on the Ground" monthly reports to Congress.*

The graph above is wildly incomplete as it only shows the number of contractors registered by the US Department of Defense (DOD). Most contractors in Afghanistan and Iraq were local citizens or third-country nationals who, for the most part, were never officially registered with DOD. No one knows the exact number of contractors, but it is likely to be several times more than represented by this graph.

Hence, taxpayers pay for services, the private sector provides the services, and the government simply acts as a very expensive and

ineffective middleman.

Private corporations often provide many services traditionally associated with government services for their employees, such as health insurance, health-care plans, and pension plans. Some corporations provide governance services as a service straight out, such as private security companies, private hospitals, and private schools. For instance, more Americans study a foreign language on the language learning app Duolingo than in the American school system.[2] Other companies have even more ambitious plans, like Virgin, SpaceX, and Amazon building space exploration facilities, or the company Sidewalk Labs, owned by Google's parent company Alphabet, building a whole new high-tech neighborhood in Toronto's waterfront area.[3]

In addition to that, many more governance services are provided peer-to-peer in communities (through neighborhood associations, religious institutions providing local safety nets for their members, various interest groups, immigrant groups, aid associations, etc.). During the European refugee crisis in 2015, Bitnation did just that: an informal online network of activists put in place the Bitnation Refugee Emergency Response (BRER) because European member states were unable to respond to the immense influx of refugees. BRER provided Emergency Blockchain IDs and Bitcoin debit cards, all paid for by Bitcoin donations from the crypto community. It was entirely set up and coded by volunteers in less than two weeks. Bitnation won the UNESCO NetExplo Award in 2017 for those efforts.[4]

Many nation states realize that jurisdictions need to be creatively applied to foster innovation and economic stimulation. There are an estimated 5,400 Special Economic Zones (SEZ) in the world where different laws apply than in their host nations.[5] In other countries, there's great regional competition, such

25 Heather Masson-Forsythe, "NSF gave Duolingo its wings!" National Science Foundation, December 22, 2022, accessed 05/03/2024, https://new.nsf.gov/science-matters/nsf-gave-duolingo-its-wings#:~:text=In%20the%20U.S.%2C%20more%20people,%2C%20Duo%2C%20needs%20little%20introduction.

26 https://www.sidewalklabs.com/

27 "The Netexplo Forum celebrated its 10th edition," UNESCO, accessed 05/03/2024. https://www.unesco.org/en/articles/netexplo-forum-celebrated-its-10th-edition.

28 Lu Fangwen, Weizeng Sun, and Jianfeng Wu, "Special Economic Zones and Human

as the United Arab Emirates (UAE) with its different emirates, Switzerland with its cantons, and the United States with its different states. This approach offers greater flexibility and liberty for the country's citizens to choose a way of life that better suits their preferences. India, one of the world's largest and fastest-growing economies, started as four private trade companies (The British East India Company, which competed with Danish, Dutch, and Portuguese trade companies) which led to the formation of the nation-states now known as India, Pakistan, and Bangladesh.[6] The government of Estonia has created the much-talked-about e-Residency program that offers governance services to anyone around the world who wants to pay for them, and others, such as the island nation of Palau, are following suit. Many nations have also started to offer 'digital nation' programs with various visa and tax liberties to attract well-off online workers to spend money in their local economies.

Afghanistan, during the time of the shah (the king)—before the Soviet invasion—was a mainly peaceful country, as nearly everywhere outside of Kabul was self-governing. In the towns and villages, it was common to see many different jurisdictions operating peacefully alongside each other, as each ethnic group followed a different set of social norms and laws. War broke out when the Soviets tried to force everyone into a one-fits-all model of governance, and the Mujahideen, the Taliban, and later on, the US-led coalition forces attempted to do the same thing. In 2011 in Libya, Susanne witnessed firsthand the relative peaceful cooperation during the early 'lawless' days of the civil war.

A personal anecdote: '*When I first came to Libya, I was surprised how well the society functioned. The only government in the eastern rebel territory was the Transitional National Council (the TNC), which was just a handful of human rights lawyers hiding out in base-*

Capital Investment: 30 Years of Evidence from China," American Economic Journal: Economic Policy 15, no. 3 (2023): 35-64.

29 William Dalrymple, The Anarchy: The East India Company, Corporate Violence, and the Pillage of an Empire (Bloomsbury Publishing, 2019), ISBN 978-1635573954.

ments and doing media statements. [7]*Initially, they had no possibility of providing any form of governance services. Instead, the civil society came together, working relentlessly to maintain a functioning society. Boy Scouts were doing traffic policing, hundreds of volunteers helped clean the streets, engineers kept the cellphone towers alive, and men paid for weapons to fight at the front from their own pocket. As time went on, the east became more solidified, the Gaddafi threat diminished, and the TNC got more and more power. That was when the different factions of society started to fight against each other to gain power, making eastern Libya vastly more dangerous than during its early 'lawless' days.'*

There are many non-geographical communities with their own set of rules that supersede those of the geographical nation they live within. The Jewish diaspora is an early example, but there are many modern examples as well, such as people working with the hacktivist group Anonymous.

THE DANGERS OF THE UNITED NATIONS WORLDVIEW

As globalization rewires our society, through global trade, transportation, and technologies, many hail the United Nations (UN) as the solution to global challenges, seeing it as a plausible post-nation-state replacement for national governments. This worldview is dangerous, as it's constructed upon the already dysfunctional Westphalian system, amplifying all its fundamental flaws.

The United Nations, established in the aftermath of World War II, was a response to the horrors of global conflict.[8] It aimed to prevent future wars, promote human rights, and foster international cooperation. The UN Charter sought to balance the principles of state sovereignty with the need for collective security.

30 Chris Stephen, "Libyan rebels struggle to set up alternative government," The Guardian, April 3, 2011.

31 Jussi M. Hanhimäki, The United Nations: A Very Short Introduction (Oxford University Press, 2008), ISBN 978-0199233163.

However, the practical application of these ideals often faces obstacles due to the diversity of cultures, interests, and power dynamics among member states.

The UN paradox lies in the tension between the global aspirations of the UN and the reality of its dependence on the cooperation of sovereign states, each with its own agenda and priorities. Its centralization is reflected in its decision-making structures, where the Security Council's permanent members hold significant influence. This concentration of power, while intended to prevent unilateral actions, has led to accusations of inaction in the face of crises. The UN is funded by member states; hence, it is incapable of independently making decisions or overruling its members, even when it is in the interest of their stated goals.

Milton Friedman once wrote, "*Nothing is so permanent as a temporary government programme*" and the UN is no exception to that rule.[9] The UN has ballooned into an organization of fifteen departments, forty thousand employees, and around US$30 billion in yearly budget. That type of mammoth-like structure is unlikely to come up with new ideas for solving existing and future challenges. Innovation mostly happens at the edges, in communities in search of real-life practical solutions to real problems, not at the complacent bureaucratic center of large organizations.

Technological advancements have always influenced governance by altering the means of communication, the spread of information, and the nature of power. The printing press revolutionized the dissemination of knowledge, challenging the dominance of religious institutions and fostering the spread of revolutionary ideas. The telegraph and later the internet transformed the speed and reach of communication, enabling global connections and instant information exchange. In the digital age, technology has given rise to new forms of governance, some of which challenge traditional centralized models. The concept of Bitnation, the first blockchain-powered network state, seeks to decentralize authority and em-

32 "The Chicago Question: What would Milton Friedman do now?" The Economist Magazine, 2012, https://www.economist.com/finance-and-economics/2012/07/28/the-chicago-question.

power self-sovereignty in the realms of governance. Blockchain's transparent and tamper-proof nature offers opportunities for secure and accountable governance mechanisms, transcending geographical boundaries. With such rapid technological advances, why do we hold onto vastly outdated governance structures?

As discussed in this chapter, the centralization of governance leads to an inability to be flexible and innovative and is often the direct cause of tensions between different groups. To further centralize governance to just a few international institutions would lead to perpetual bloodshed, as people do not like to be told by a faceless bureaucracy far away how to live their lives. And the UN is highly unlikely to be able to adapt to challenges because of its large and ineffective bureaucracy further muddled by compromises with member states.

A CONCRETE PROPOSAL TO THE INTERNATIONAL GOVERNANCE PRACTITIONER COMMUNITY: MAKE A LEGAL FRAMEWORK FOR "CREATIVE GOVERNANCE ZONES" BOTH ONLINE AND OFFLINE

We believe that many people in power today come from the generation who felt a strong buy-in into the current system; they felt like they were a part of shaping it. From the post-WWII boom to the end of colonialism as we know it and the Cold War, to the creation of the European Union and the early days of the UN and NATO, it came from a good place. They genuinely wanted to create a better, safer, and fairer world for everyone. They choose people to fill their ranks who had the same worldview as they did, creating a great uniformity of thoughts and policies. They tried to mitigate that through 'diversity hires' policies, which created a different sort of problem by promoting people based on criteria other than talent. However, it still did not solve the problem of uniformity because the diversity hires agreed with everything those in power promoted in order to get hired. The problem of the lack of challenge to the system and the lack of freedom to innovate and find creative solutions cannot be solved simply by having a person of different skin color or gender saying the same thing as the powerholders. Innovation does not come from government innovation labs or a banks 'Innovator Incubator.' Innovation happens at the edges, where people find real solutions to real problems, and those solutions might not match the exact color prism through which those in power wish to view the world. However, it is essential that these solutions are allowed to exist for the progress and betterment of humanity and our societies. We, too, want to build our own society and political solutions, just like the previous generation did in many ways. It is also a way to honor their values: feeling like we have an actual role to play in something that is happening and staying engaged in how society functions, changes, and improves. Why wouldn't they want us to have the same experience they did? Are they so arrogant that they think the solutions they produced or worked on are eternally the best and not meant to evolve and change with time? That sort of arrogance does not look good on a community that prides itself on openness and fairness for all.

Hence, here's a proposal: Let everyone create their own nations, like we create our own companies, our own software. Many will fail; some will work. Either way, we will learn, iterate, and improve—something that current nation-states are too large and inflexible to do effectively. Whether it's a geographical micronation like Liberland or Liberstad, or attempted floating cities, or a network-state such as Bitnation, we need the liberty to help create new solutions for our mutual society. Creating a legal framework around that, recognizing a quasi-political legitimacy of such attempts, would help us do so. I propose the name "Creative Governance Zones"; these zones would give a semi-political autonomy, and could include rights to legally provide governance services, which would receive recognition as such from nation-states and supra-national governance bodies, examples being IDs, notary services, and dispute resolution services.

That would be a good first step to give a new generation of innovators, entrepreneurs, and technologists a feeling of buy-in into the development of society, and the incentive to invest time and energy to invent creative solutions without hindrance from a stale and inflexible system that has effectively trapped billions of people in a geographical apartheid. Change will happen anyway, inevitably. People can choose to embrace it and work with it peacefully. Or, they can try to use government restrictions to stop it and take their arrogance with them all the way to the grave while fostering a more apathetic and dysfunctional society in the process, ultimately leading to a dangerous implosion.

CHAPTER FIVE

THE IDENTITY CONUNDRUM: PERSONHOOD IN A HYPER-CONNECTED AGE

UNRAVELING IDENTITY: UNDERSTANDING THE COMPLEXITIES AND IMPLICATIONS OF IDENTITY IN A DIGITAL WORLD

Individual identities are the parameters by which a person defines themselves in society. They can be based on national identity, identity within a social group, identity within a family unit, professional identity, ethnicity-based or religious identity, or any other cultural identity. Often, all these identities are meshed together. Individual identity forms gradually as religions emphasize individuals' responsibilities for ethical behavior, the industrial revolution makes work more specialized, family units gradually become smaller, and the study of psychology becomes mainstream in society.

Religious, national, ethnic, and family identities are group or collective identities, while individual identities might not be based on anything more than what that person has done for themselves or how they see themselves. There are also internal and external identities. Internal identity is entirely how you see yourself, while external identity is often imposed upon you by society, like the caste systems of India, which then becomes internalized. The question is, will society let individuals have an individual (self-defined) identity or will it be imposed on people like it was in Nazi Germany? If identity is imposed by society, such as in the caste system of India, is there any individual in that instance? In the USA, there is debate around gender identity. Trans people want others to accept their gender identity, but there are many who refuse to address them by their chosen identity, so instead, they receive a collective identity that they did not choose or believe in. It is worth noting that imposed identity is distinct from the identity the person internalizes. And as we develop newer identities, like cyborg or transhumanist, these may be mocked or not accepted as real identities at all.

Today, it can, at times, seem like everything revolves around identity. From the left side of the political spectrum to the right side, they accuse each other of engaging in identity politics, to the point where 'identity' has become a toxic word in popular discourse. And it's easy to see why; in a world where life is fluid, where we have a near-infinite number of opportunities to connect with people from all countries, from all walks of life, where everything is interconnected and we're told that we can be whoever and whatever we want, the need to define oneself has never been greater, or more challenging. As of today, all humans agree on one thing: we are all Homo sapiens. But that has not always been the case, and it is very unlikely it will always be the case. Humans used to be divided between several distinct species until Homo sapiens became the dominant—and eventually, the only—human race. However, as humans are fusing with technology, it is reasonable to think that the time of Homo sapiens singular form or even superiority will not last much longer. When humans can indefinitely extend their lifetime, replace virtually every organ in the body, modify their genome, eradicate sickness, and grow children in artificial wombs, a new form of human species will emerge. The change is likely to happen unevenly throughout society, as some will not want to change, and others will not be able to afford it. That will certainly turn the identity debate on its head like never before.

However, in this chapter, we will focus on more pressing issues we are facing already today when it comes to identity creation and management. When speaking about identity in the context of governance, we must talk about identity documents (IDs), such as ID cards, passports, driving licenses, and student cards. There is no worldwide identification system currently in place today, though passports are the universally most accepted form of ID. Currently, most people get their ID from their nation-state government, but the quality of such documents varies depending on what country issues them, and there is a large market for counterfeit IDs. This adds to the miseries caused by the global Westphalian geographical apartheid.

It's also unfit for the many hundreds of millions of people who do not live in their country of citizenship, such as expatriates, digital nomads, and refugees, not to mention people who are stateless and have no valid passport or nationality at all. It might be hard to obtain the ID from your country of citizenship when residing abroad, and it might be less useful as a document, entitling you to fewer rights in the country of your residence.

That problem is likely to grow significantly over the coming decades, mainly due to the growing number of refugees worldwide. Australian think tank IEP predicts that at least 1.2 billion people could be displaced by climate-related events by 2050.[1] Many are likely to remain without a fixed country of residency for many years and will need international protection beyond what the nation-states are capable or willing to provide with their current means. A World Citizen ID would be a critical first step in defining and empowering such global rights.

---

32 Jocelyn Wyatt, "How climate change is changing the definition of 'refugee'," 2023, https://www.oecd-forum.org/posts/how-climate-change-is-changing-the-definition-of-refugee#:~:text=The%20report%20also%20urges%20us,response%20to%20the%20climate%20crisis.

THE QUEST FOR A WORLD ID: INVESTIGATING THE FEASIBILITY AND CHALLENGES OF IMPLEMENTING A GLOBAL IDENTITY RECOGNITION SYSTEM

Having a unique identifier is important for several governance purposes, for instance:

- 1 person = 1 vote (in voting systems).
- 1 person = 1 financial access (e.g., in case of emergency aid packages when resources are limited).
- 1 person = 1 specific right to access certain governance zones and services.
- 1 person = 1 signature on contracts such as trade contracts, loan contracts, rental contracts, and employment contracts.

To guarantee that you are a unique human (as opposed to a deepfake, an AI, etc.)

It's not always important to have a unique identifier if there's no point in gaming a system. A person could, for instance, have many different wallet addresses or online aliases for different purposes. But in this chapter, we'll focus on the one person = one identifier system.

In 2014, Bitnation recorded the world's first World Citizenship ID together with Chris Ellis on the Bitcoin blockchain.

"The goal of this project is to learn and layout a simple process for anyone in the world to create their own Private Passport Service that can be used to validate and prove the existence of other persons using nothing but available tools. We will prefer open source where available and we will draw on cryptographic tools like Public Private Key Cryptography (PGP) and blockchain technology in the form of Bitcoin. By doing this we aim to give people across the world the ability to grant one another Global Citizenship by virtue of their being witnessed in space and in time.

This witnessing can be documented with photography and video, that content can be signed with PGP signatures, hashed and timestamped. It can then be joined with Social Network Validation services like Keybase.io before being plugged into more dynamic reputation systems."

— **Christopher Ellis**, London, UK, 2014

Figure 22: ***The first Blockchain based World Citizen ID Card***

Since then, there have been a myriad of attempts to create a global ID, though none have found any mainstream traction yet. The latest stab at a blockchain ID is called WorldCoin, launched by the nonprofit Tools for Humanity, which was funded by Sam Altman (the founder of OpenAI/ChatGPT) and Alex Blania.[2] The project has generated a significant deal of controversy among both the crypto community and various governments because of its use of iris scan technology through a scanning device the organization calls "orbs." At the time of writing, over 5 Million persons have registered as World Citizens on their mobile phone app. WorldCoin has been advertised as a possible attempt to provide some form of basic income through distributing their token WorldCoin (WLD) to

34 https://worldcoin.org/.

those who are identified or enable others such as nation-state governments to do so. Time will tell if WorldCoin or some other project or protocol gains the most adoption and usage.

The reason many software development teams are working on blockchain identity is because of the tamperproof nature of the blockchain. It is secure, and if done correctly, pseudonymous. However, the danger also lies in its resilience. The information is stored on the chain forever, which can be dangerous for a person. It might compromise the security of that person if for instance, the person is facing prosecution by a group of people because of their political or religious affiliations. Although some chains do disappear if there is not anyone to maintain nodes, with mainstream chains like Bitcoin or Ethereum, that risk is practically nonexistent. Another plausible threat is the coming quantum computing abilities. Many developers are working on quantum-safe blockchain technology, but it has not been tested yet. Hence, there is a risk that quantum computers will crack the blockchain encryption and spread or wipe out all the data stored on the chain.

Imagine if you are a Rohingya in Myanmar living through the world's biggest ongoing genocide, and you have to flee your village for fear of being killed, raped, or imprisoned because of your ethnicity and religion.[3] You flee to a nearby town where you can get a bus to a safer region or closer to a border crossing into Bangladesh. Imagine that a blockchain ID, government-issued or not, has become the standard for everything, even something as mundane as a bus ride. Imagine that even though "you control what data you want to show," as most blockchain ID projects promise, in order to sign up for an aid project for underprivileged populations to receive a badly needed malaria medication, you had to use a third-party app that matches your blockchain ID to your data in a public database, with, for instance, your place of birth and your ethnicity. Hence, imagine being scanned through your iris or some other biometric means at the bus stop (e.g., facial recognition, fingerprints) and not only being denied getting on

35 Human Rights Watch, https://www.hrw.org/tag/rohingya#:~:text=In%20Myanmar%2C%20about%20600%2C000%20Rohingya,coup%20and%20brutal%20nationwide%20crackdown.

the bus, but also being directly reported to the authorities as a plausible refugee. Then, a couple of policemen show up and shoot you on the spot. That is the danger of biometrics on the blockchain. You cannot delete the data, and you cannot throw away your ID because it is a part of your body, unless you go through some extremely sophisticated plastic operations or amputations. That is extremely dangerous for many people in precarious situations, and there is no way to know who will be in that sort of situation next time. It could be any of us for any number of reasons we have yet to imagine.

Many favor anonymous and pseudonymous identities to increase individual security and control over their identity. An anonymous identity simply proves that you are a unique human being, nothing else. A pseudonymous identity, however, follows you around and can be attached to a reputation, whether formal or informal. An example is online vendors who often do not use their real names while selling but are still subject to reputation scores by their clients, which determine their future success (or lack thereof) as vendors. It still offers a degree of privacy protection while having many more functions than simply a unique identifier.

PROOF OF PERSONHOOD: DELVING INTO THE CONCEPT, EXPLORING ITS SIGNIFICANCE, AND EXAMINING THE ROLE IT PLAYS IN MODERN DIGITAL IDENTITIES

"Proof of Personhood" is a concept in the domain of digital identities and decentralized systems that seeks to verify the uniqueness of individual human participants without necessarily revealing their identity. The goal is to ensure that each person in a system corresponds to a unique, real-world human being without compromising their privacy. For use cases such as financial transactions, voting, or any scenario where the participant might have a reasonable expectation of privacy, they should not be required to face the dilemma of giving up privacy to establish their identity. To achieve these goals within the context of a decentralized network, there must be a mechanism for personhood verification, and proof of personhood is a concept designed as a method of achieving such verification.

The Significance of Proof of Personhood: The most obvious example of proof of personhood's significance is in preventing Sybil attacks. In many decentralized systems, participants can gain undue influence by creating many pseudo-anonymous identities. This is called a Sybil attack. If a system can ensure that each participant represents a unique human, it becomes much harder to conduct these kinds of attacks.

Fair distribution is also important to many. Whether the distribution is about something like universal basic income (UBI) or, in the cryptocurrency space, fair distribution is numerically measurable by a number called the Gini coefficient. In systems with very unfair distribution, you will have this unfairness represented in the Gini coefficient.[4] In proof of personhood, the distributions can be given out on a per-human basis rather than a per-account basis.

Voting systems require proof of personhood for the same sorts of reasons. In voting, it's supposed to be one person, one vote. It is important to note that Satoshi Nakamoto mentioned in the Bitcoin whitepaper the concept of "one-CPU-one-vote" and "one-IPAddress-one-vote." It was pretty obvious that for Bitcoin to work, they needed to invent "Proof of Work" as a means to make Sybil attacks economically expensive. A new CPU has an economic cost greater than that of a new IP address. For these reasons, it would be economically costly for a single entity to gain control of over 51 percent of the Bitcoin network, which, if you think about it, is similar to how a corporation works: as long as you maintain 51 percent control, you keep ownership.

36 Cesar A. Hidalgo and Ricardo Hausmann, eds., Economic Inequality: New Methods for New Trends (Palgrave Macmillan, 2018), ISBN 978-3319904241.

Traditional identity verification systems might require government IDs or other personally identifiable information. In contrast, some methods of proof of personhood can verify a user's uniqueness without revealing who they specifically are. It is important to note that some of these methods were invented before the emergence of deepfake technology.

Anti-spam and rate-limiting are possible if we can determine that a request or output was generated by a unique human. This can disrupt spammy behaviors and reduce the noise. This is important for content filtering, and we can see this on networks like X (formerly Twitter), where bots threaten to overrun the network with state-run trolls.

HOW PROOF OF PERSONHOOD APPROACHES SOLVING THE PROBLEM OF PERSONHOOD VERIFICATION:

Several solutions were proposed to solve the problem of "proof of personhood." One of these solutions is called "pseudonym pairs".[5] The concept of pseudonym pairs evolved from a solution called "pseudonym parties".[6] Bryan Ford discusses in his 2007 paper how this solution can provide an alternative method of establishing the "one person, one persona rule" currently being attempted by WorldCoin, as mentioned above.

In saying all of this, we have to accept that proof of personhood comes with challenges that are currently difficult to address. Some methods for proof of personhood, such as the "Democracy Earth" approach, require physical gatherings. These are difficult to scale and are prohibitive for those in remote areas. In addition, some methods rely on biometric data, like fingerprints and facial

37 Johan Nygren, "Pseudonym Pairs: A foundation for proof-of-personhood in the web 3.0 jurisdiction," (2019).

38 Bryan Ford, "Pseudonym Parties: An Offline Foundation for Online Accountability," (March 27, 2007).

recognition. These biometric methods raise privacy concerns: Where will this data be stored, and who or what could be allowed to access it, like in the case of the Rohingya refugee? It is also possible to game these systems, and there is the risk that people will be able to bypass the security measures and falsely represent themselves as multiple unique humans.

Most importantly, if the system relies on advanced technologies, then we have the risk that some will be excluded. Not everyone will have access to the same level of technology. This presents a risk that some may be excluded from having an identity because they are not in a high-tech society. As a result, a lot of the challenges of proof of personhood remains unsolved, and will be one of the most important and daunting obstacles to overcome in the next few years.

CHAPTER SIX

WHISPERINGS OF ACCELERATION: MOORE'S LAW AND TECHNOLOGICAL ACCELERATIONISM

"Accelerationism" is a term created by science-fiction author Roger Zelazny for use in his book titled Lord of Light.[1] Lord of Light, in brief, includes a specific character named Sam, who has the idea that he could introduce his religion (Buddhism) to the exoplanet.

For a deeper understanding, Sam is from a crew that populated the new planet, but he chose to rebel and kickstart a revolution. Sam drew inspiration from Buddhism and embodies the accelerationist spirit. By leveraging the very technologies and religious structures that the pseudo-deities use for control, he accelerates societal evolution, aiming to free society from technological and spiritual oppression. Sam wanted to distribute technology to the masses rather than keep the technology among the elites like himself.

39 Roger Zelazny, Lord of Light, (1967).

Accelerationism is a philosophical and socio-political concept that advocates for expediting technological and societal advancements to induce transformative change. It posits that intensifying the processes of capitalism and technology, rather than resisting them, can lead to radical social upheaval and a post-capitalist future. Some adherents believe this acceleration will destabilize the system from within, while others see it as a means to achieve a technologically driven utopia.

Figure 23: ***Roger Zelazny, Lord of Light***

Many transhumanists base their futurist projections on Moore's law. Moore's law is expressed as a formula where n = *the number of years to develop a new microprocessor divided by 2*. In other words, the number of transistors is predicted to double because the time to develop a microprocessor decreases every two years.

WHAT ARE THE IMPLICATIONS OF MOORE'S LAW?

With more transistors for cheaper over time, we get a continued reduction in the cost of computing. This is almost like a deflationary force generated by Moore's law, which, when combined with AI, could result in the disruption of millions of jobs.

Bitcoin is based on and benefits from the fact that Moore's law occurs, which, at least in our opinion, was an assumption in the design of Proof of Work. Proof of Work relies on the fact that computing will become more efficient, which would imply that eventually, even in the worst-case scenario, Proof of Work will become energy efficient due to the development of ASIC technology. ASIC, which stands for 'application-specific integrated circuit,' is a kind of hardware chipset that specializes in a specific algorithm or purpose. Imagine you have a box of special Lego pieces that you can only use to build one specific thing, like a spaceship. An ASIC is like that box of Legos. It's a tiny part inside electronics, like a phone or a computer, that's made to do one job really, really well. Just like you can build a great spaceship with that special Lego set, the ASIC is really good at its one job, but it can't do anything else.

Georgi Karov, a lawyer, a deep philosophical thinker, and the author of the book "*The Force in Bitcoin, labels Bitcoin an "Algorithmic Commodity."*[2] He highlights these facts:

40 Georgi Karov, The Force in Bitcoin (November 11, 2021), ISBN: 979-8670620871.

Bitcoin's total supply is algorithmically determined.

Bitcoin's scarcity is programmatically set to grow over time.

Bitcoin's destruction (known as burning or losing private keys) is not capped.

Bitcoin's guaranteed scarcity excites higher demand.

The Bitcoin reward halving event, which takes place every four years, is a reduction in the emission rate of new supply. We can then rely on the laws of economics to understand that Bitcoin price is based on the formula

$P=f(QD,QS)$, where:

P is the price of the good or service.

QD is the quantity demanded at that price.

QS is the quantity supplied at that price.

Equilibrium occurs when $QD=QS$, meaning the quantity of the product that consumers are willing to purchase at a particular price equals the quantity that producers are willing to supply. At this point, the market price is stable.

Translated simply, it means the price of a good or service is a function of the quantity demanded at that price and the quantity supplied at that price reaching equilibrium. When supply matches demand at a particular price, you achieve price stability. Bitcoin supply increase over time is halved at a rate of every four years, which resets the equilibrium point—at least, theoretically.

Alex Wissner-Gross, a computer scientist who taught at Harvard and MIT, proposed a formula related to intelligence, which he defines as a force to maximize future freedom of action and avoid constraints in its own future: $F=T\nabla S_\tau$. One of his flagship ideas includes intelligence as a property.

In this context:

F represents a force that models intelligence.

T is a variable representing temperature or a similar scaling factor.

$\nabla S\tau$ is the gradient of entropy over time (τ).

This formulation suggests that intelligent behavior could be universally defined as actions that maximize future freedom of action or options. It's a thermodynamic approach to understanding intelligence, proposing that the essence of intelligent behavior is to maximize potential future pathways or options.

Another interpretation of this same formula, proposed by Karl Friston, is called the "Free Energy Formula": $F=T\nabla S\tau$. In the context of bio-inspired AI, this principle can be thought of as a metaphor for how intelligent systems (both biological and artificial) maintain their internal structure and function in the face of a constantly changing environment. They do this by minimizing 'free energy,' which can be understood as minimizing surprise or uncertainty about their sensory inputs. In this equation, F represents free energy, a concept borrowed from thermodynamics and statistical physics, while $T\nabla S\tau$ represents the gradient of entropy over time (τ). This term encapsulates the idea that living systems resist a natural tendency towards disorder (entropy). This principle is particularly influential in theories about how the brain functions and has implications for developing AI that more closely mimics human cognition and adaptation.

ChatGPT relies on Moore's law as well. Large language model AI leverages the fact that the cost of computing is predicted to keep getting cheaper. In the case of ChatGPT, it is the means by which they can scale by throwing more processing power at it (right now measuring at around 800 petaflops).

The father of ChatGPT, Sam Altman (not to be confused with Sam from Lord of Light), modified the idea of Moore's law with a new statement that, in our opinion, is pertinent:

"The amount of intelligence in the universe doubles every eighteen months." Whether or not Altman's Law becomes true is still up for debate because the amount of intelligence in the universe is unknown.

Intelligence Doubling (based on Altman's speculation)

$$I(t) = I\_0 \times 2\{t/18\}$$ *Where:*

- I(t) represents the amount of intelligence in the universe at time t, and
- I0 is the initial intelligence at time t=0.

Consider the interactions that occur between these formulas. The formula for Moore's law is talking about the foundational substrate from which AI is to double. What Sam Altman is doing here is drawing an analogy between Moore's law and intelligence doubling. We do not yet see evidence in data supporting a correlation between Moore's law and growth of intelligence in the universe, so we must label his statement entirely speculative at this time.

Altman is making a forward-looking statement about the amount of intelligence in the universe (unknown), the trajectory of AI, and its implications, rooted in an optimistic view of technological progress.

Let's assume for a moment that this speculation is true. Given these inputs as a basis of our assumptions, we can produce a formula: $C(t)=UBIr \times AIm \times I(t)$

In this formula:

I(t) = Total Intelligence at time t. This includes both human and machine intelligence. The "intelligence explosion" idea suggests that I(t) grows exponentially, primarily driven by AI advancements.

AI_m = Multiplier due to AI. This captures the rate at which AI contributes to intelligence growth, and it's assumed to be greater than 1 (indicating exponential growth).

UBI_r = Rate at which universal basic income (UBI) is funded by AI. As AI drives more economic value, a portion of that value is channeled into UBI.

C(t) = Citizen's or World Dividend at time t. This is a crypto token whose value is tied to I(t) and the economic value derived from it.

Proposed Formula:

$C(t)=UBIr\times AIm\times I(t)$

This formula encapsulates the idea that the Citizen's Dividend (C) at any given time (t) is a function of:

the total intelligence at that time I(t),

the rate at which AI is driving that intelligence growth (AI_m), and

the rate at which the economic value derived from this intelligence is being channeled into UBI (UBI_r).

This formula is important to make note of for Chapter 9: "Reflections on Universal Basic Income and the Artificial Intelligence Tango." The main assumption of that chapter is that the profit from any potential future intelligence explosion as predicted by AI researchers can be captured and distributed to humanity as a citizen's dividend. This could either be done in a decentralized manner using cryptocurrency or in any manner of centralized organizations. Chapter 9 goes into the nuts and bolts of the idea and how it can possibly be implemented.

CHAPTER SEVEN

42 SHADES OF THE EVER-SHIFTING REPUTATION ECONOMY

VARIATIONS OF THE REPUTATION ECONOMY

Humanity has set itself apart from other species on planet Earth through its ability to discover and build extraordinary things, from fire and the wheel to electricity, cancer treatments, and space flights. This has been possible through humans' ability to communicate and cooperate across many stakeholders, through sharing information, exchanging goods and services, and incentivizing positive competition rather than mutual destruction. In our interconnected world, it is logical to work on a large scale with global reputation systems, both for individuals and for enterprises, institutions, and nations.

Reputation systems have been integral to social coordination, even before the advent of modern economies. Since the dawn of humanity, each tribe member, family member, and so on has been known and appreciated by other members of their tribe or family for their attributes, virtues, and past actions (which is sometimes referred to as the 'big grandma' society, in comparison to the more centralized notion of a 'big brother' society). Reputation systems help communities incentivize desired behaviors and discourage harmful ones. With the emergence of blockchain technology, creating decentralized reputation systems based on tokenized rewards has become more feasible than ever, and potentially more scalable than ever. But how exactly do these systems influence behavior? To understand this, we need to consider the "spectrum of behaviors."

The Spectrum of Behaviors – An Overview: The spectrum of behaviors is a concept that posits a range of possible actions an individual can take in response to a given situation. These potential behaviors are not uniformly likely; instead, their probabilities are influenced by the incentive map in the environment. This perspective finds roots in B.F. Skinner's behaviorism theory, which suggests that environmental stimuli, particularly rewards and punishments, condition and shape our behaviors.[1] Imagine living in a world where everyone wears a digital badge displaying their reputation. This badge could change color based on the person's actions—green for kind acts, red for unkind ones—as one of many ways to possibly implement the reputation system. We have to be careful in the design of this kind of reputation system because it can enable retributive troll behaviors if done wrong. In Bitnation, we designed a reputation system that was very careful to take into account the potential perils of peer-to-peer reputation systems. In order to avoid a dystopian society, we designed it so that the network only rewarded good behavior rather than punish bad behaviors.

The Ethereum blockchain has a mechanism for rewards and slashing, which in my interpretation can be likened to a virtual Skinner box. A Skinner box is a concept derived from B.F. Skinner's experiments in operant conditioning. Here's how this analogy works:

Rewards in Ethereum: In the context of Ethereum, validators (participants in the network who validate transactions and add new blocks to the block-

41 B.F. Skinner, Science and Human Behavior (Free Press, 1953), ISBN: 978-0029290408.

chain) receive rewards for their contributions. This is akin to the positive reinforcement in a Skinner box, where an animal is rewarded for a desired behavior (e.g., pressing a lever).

Slashing in Ethereum: Slashing is a penalty mechanism in Ethereum where a portion of a validator's stake is taken away if they act maliciously or fail to fulfill their responsibilities. This is similar to the concept of punishment in the Skinner box, where undesirable behavior is discouraged through negative consequences.

Behavioral Conditioning: Just like the Skinner box conditions animals to perform certain actions through rewards and punishments, Ethereum's reward and slashing mechanisms incentivize validators to act in the best interest of the network (the community, society, and so on). Validators are encouraged to act honestly and efficiently to gain rewards and avoid penalties, thus ensuring the integrity and security of the network.

Feedback Loop: The continuous process of receiving rewards and avoiding slashing creates a feedback loop for validators. They are constantly adapting their behavior to maximize rewards and minimize penalties, akin to how animals in a Skinner box learn to modify their behavior over time based on the consequences they experience.

A baby is not inherently good or evil. A newborn baby, in fact, is incapable of evil behavior. The range of behaviors of a baby is very limited compared to the range of behaviors of an adult. To visualize the concept of a behavioral spectrum, just imagine at first a baby, which has a set of common behaviors shared with all babies, such as crying, but it is also known that there are many behaviors a baby cannot engage in, which is why a baby is considered harmless. As a baby becomes a toddler, their range of possible behaviors increases along with their intelligence level, and eventually, every human develops a concept of reward and punishment, action and consequence, and over time, these basics become systematized into belief systems, understandings of right and wrong. Some elements of these belief systems are based on the innate nature of the child, but most of it seems to be learned.

The total spectrum of possible behavior is a probability distribution. For each developmental stage, we define a set of possible behaviors. We assign a probability distribution over the set of behaviors for each stage. This is represented by a function Pstage(b), where b is a behavior in the set Bstage. This function returns the probability of behavior b occurring in the given developmental stage. The environment affects these probabilities. The environment is represented as a vector E=(e1,e2,...,en). The modified probability distribution, which takes the environment into account, is represented as Pstage(b|E). As an individual progresses through different stages, the set of behaviors and their probabilities change. This can be modeled with a transition function T(Bstage1,E)→Bstage2. The overall behavioral spectrum is a function S(age,E) that maps age and environmental factors to a probability distribution over behaviors. It yields Pstage(b|E) for the developmental stage associated with a given age. The result is S(age,E)=Pstage(b|E). Finally, consider individual differences or individual predispositions as a factor P, modifying the probability distributions to Pstage(b|E,P).

- Defining the Behavioral Set: $B_{\text{baby}} = \{\text{crying, laughing, sleeping,...}\}$
- Probability Distributions: $P_{\text{stage}}(b)$ |
- Influence of Environment: $P_{\text{stage}}(b \mid E)$
- Developmental Progression: $T(B_{\text{stage1}}, E) \rightarrow B_{\text{stage2}}$
- Behavioral Spectrum Function: $S(\text{age}, E) = P_{\text{stage}}(b \mid E)$
- Individual Differences: $P_{\text{stage}}(b \mid E,P)$

Note that this is a very rough outline of the concept. The general understanding is that behaviors are probabilistic in nature. The probability distribution allows us to understand this probabilistic nature mathematically. The unpredictable nature of the environment and its influence is a major factor in changing the probabilities of different behaviors that can emerge from a person. The developmental stage of a person is also a major factor in determining which behaviors could emerge. The behavior spectrum function is what takes all of this into account, and when you also factor in individual differences/predispositions, you can have a mathematical understanding of human behavior.

The assumption in this understanding is that there is no good or evil person. There are good and evil behaviors based on what the judger of the behavior determines to be good or evil and/or based on the consequences of the behavior. This would imply that no one can be said to be inherently anything, as any behavior that can exist could emerge from anyone, but the probabilities of their emergence differ according to the factors included in the mathematical formula. For example, a baby cannot ever be a serial killer, or ride a bike, or engage in any behavior that most of humanity could recognize as evil, but as the baby continues to develop and its cognitive powers improve, the spectrum of potential helpful or harmful behaviors grows. But what maximizes the positive behaviors and minimizes the harmful behaviors? One way is through operant conditioning and incentive mapping.

Operant Conditioning and Incentive Mapping: Just as mice can find their way through a complex maze by following the scent of cheese, humans can navigate through the labyrinth of social norms and systems, guided by the metaphorical smell of 'cheese'—the incentives offered or penalties enforced by their environments. Our behavioral spectrum narrows or widens in response to these incentives, a process akin to Skinner's concept of operant conditioning. The shaping of behaviors in correlation to the incentive map forms the crux of operant conditioning theory. Consider a mouse in a maze, with a delicious piece of cheese waiting at the end. The mouse quickly learns to avoid dead ends and take the right paths to the cheese. Similarly, in life, we tend to act in ways that lead us to our 'cheese'—the best results. Remember that even with good incentives, some individual differences exist, where some people might have predispositions, but in general, there is some reward or sense of winning that each person welcomes.

The process of operant conditioning used by a mouse to learn its way through a maze has notable similarities to the methods used by an AI, particularly those employing reinforcement learning algorithms, to navigate through a virtual maze. Here's how they are similar:

Trial and Error Learning: Both a mouse and an AI use trial and error to navigate the maze. The mouse explores different paths, learning from its experiences, much like an AI algorithm tries various actions and learns from their outcomes.

Feedback and Adaptation: In operant conditioning, a mouse receives feedback (like finding food at the end of the maze or hitting a dead end). Similarly, AI receives feedback through rewards or penalties based on its actions. Both use this feedback to adapt their future behavior—the mouse learns which paths lead to rewards, while the AI adjusts its strategy to maximize rewards.

Reinforcement Signals: For a mouse, rewards (like food) and punishments (like a mild shock or a dead end) serve as reinforcement signals. In AI, rewards are predefined by the programmers (e.g., a positive number for reaching the goal, negative for hitting a wall). Both the mouse and the AI use these signals to guide their learning process.

Incremental Learning: Both the mouse and the AI learn incrementally. The mouse does not understand the entire layout of the maze instantly but gradually learns from repeated attempts. Likewise, the AI starts with little to no knowledge about the maze and incrementally improves its understanding and strategy through repeated trials.

Behavioral Modification: As the mouse runs through the maze multiple times, it modifies its behavior, avoiding paths that lead to dead ends and favoring those leading to rewards. Similarly, the AI algorithm modifies its decision-making process, learning to choose actions that increase the likelihood of reaching the goal.

Memory and Experience: Both the mouse and the AI rely on memory and past experiences to navigate more effectively. The mouse remembers which paths were successful, while the AI uses its history of actions and outcomes to inform future decisions.

In essence, the way a mouse uses operant conditioning to learn its way through a maze is conceptually similar to how an AI, particularly one using reinforcement learning, navigates a virtual maze. Both involve learning from feedback, adapting behavior based on past experiences, and using trial and error to find the most efficient path to a goal.

INCENTIVE MAPPING

Incentive Mapping in a Reputation Economy: In a reputation economy, the incentive map is primarily determined by the perceived benefits of having a good reputation and the detriments of a bad one. The more valuable a good reputation is within the community, the stronger the 'scent of the cheese' becomes. Positive behaviors that earn esteem from peers are thus incentivized, whereas behaviors that could damage one's reputation are discouraged.

Imagine the reputation badges in our story. People who are often green, indicating kind behaviors, are rewarded with better treatment by others. Those with red badges are usually avoided or given fewer opportunities. This is how a reputation economy would influence our decisions and actions.

Implications of the Spectrum of Behaviors on the Reputation Economy: By adjusting the incentives in the reputation economy, we can influence the spectrum of behaviors to align more closely with the values and goals of the community. However, it's also crucial to understand that there are limits to this approach. Incentives can drive behavior to a great extent, but they are not the only factors at play. Personal beliefs, intrinsic motivations, and individual circumstances also have a significant role.

The color-changing badges in our story would make us think more about our actions. We'd want to avoid the red badge, so we'd be less likely to act in ways that harm others. We also want to encourage kindness and considerate behavior in other ways because it's also important to remember that our actions shouldn't be driven by badges alone. For these reasons, it's important to find the right balance in the search for harmony.

BALANCING THE SPECTRUM

While reputation economies, underpinned by the spectrum of behaviors, have the potential to drive desirable actions and discourage harmful ones, it is vital to strike a balance. Over-reliance on extrinsic motivators could overshadow intrinsic motivations and may not cultivate a genuine commitment to community values. The challenge, therefore, is to create a reputation economy that encourages a behavioral spectrum conducive to social well-being and cohesion, while respecting individual autonomy and diversity.

NATURE'S BEHAVIORAL COORDINATION

Pheromones: Pheromones are chemical signals used by many animal species to communicate and coordinate behaviors. They are secreted to convey a variety of messages and invoke specific reactions within the same species. For example, ants use pheromones to create trails leading to a food source. These scent trails are strengthened by each ant that follows them, helping others in the colony to find the food. The pheromone trail effectively serves as a behavioral incentive, guiding the ants along the most rewarding path. Picture a busy colony of ants. Each ant diligently follows a specific path, guided by an invisible pheromone trail. The more ants that follow the trail, the stronger the scent becomes, helping others in the colony to locate the food source.

The Ant Colony Optimization algorithm (ACO), often referred to as the "ant hill" algorithm in computer science, is a probabilistic technique for solving computational problems which can be reduced to finding good paths through graphs. This algorithm is inspired by the behavior of ants searching for food. Ian Clarke implemented an algorithm similar to this for Freenet, which was one of the earliest projects creating the dark web. Here's how it typically works:

Modeling Ant Behavior: The algorithm models the behavior of ants laying down pheromones on paths between their colony and food sources. In nature, ants initially explore the area around their nest randomly. As they find food, they return to the nest while laying down a pheromone trail.

Pheromone Trails: Other ants perceive these pheromone trails. The more pheromone a path has, the more likely other ants are to follow it. Over time, the shortest path between the nest and the food source becomes the most traveled route, as the pheromone on this path is reinforced constantly.

Simulation in ACO: In the ACO algorithm, a number of "artificial ants" simulate this behavior. They explore possible solutions to the problem, laying down and following "pheromones" (i.e., probabilistic markers) that represent how good a particular solution is.

Pheromone Evaporation: To avoid the convergence to a locally optimal solution and to explore new paths, the algorithm implements pheromone evaporation. This mimics the natural process of pheromone evaporation over time, which prevents the paths from becoming overly attractive and thus allows the exploration of alternative paths.

Updating Pheromones: After all ants have completed a tour or a cycle (i.e., found a solution), the pheromones are updated. Paths that were part of better solutions receive more pheromone, while those that were part of poorer solutions receive less. This guides the next generation of ants toward these better solutions.

Iterative Process: The process is repeated for a number of iterations or until a satisfactory solution is found. With each iteration, the pheromone trails guide the ants to explore solutions that are close to the best ones found in previous iterations.

Problem-Specific Adaptations: The basic idea of ACO can be adapted to a variety of optimization problems. The way ants lay down pheromones and the rules they use to move can be tweaked according to the specific problem being solved, such as traveling salesman, job scheduling, or network routing problems.

Digital Pheromones: Reputation points or tokens in a blockchain-based reputation economy functioning as digital reputation systems and tokenomics analogous to pheromones in the animal kingdom could be called "digital pheromones." These digital markers serve a similar function to the ants' pheromones: communicating information and shaping behaviors within a given ecosystem. In a blockchain environment, these tokens—called non-fungible tokens (NFTs)—can represent reputation, earned or lost based on an entity's actions, serving as incentives that can encourage or discourage behaviors.

Each positive interaction or contribution to the community might add tokens to a user's digital wallet, similar to the strengthening of a pheromone trail leading to food. Conversely, negative actions might result in token deductions, akin to a fading scent trail. This system thus creates a kind of behavioral map, with digital pheromones acting as signposts pointing toward the actions that are most valued in the community. Imagine the digital badges from our earlier example, now represented by tokens. Each time an entity performs a kind act, they receive tokens in their digital wallet. Others in the community can see the number of tokens each entity has, just as ants can detect the strength of a pheromone trail. The more tokens an entity has, the stronger their reputation, guiding others to adopt similar behaviors.

We see rating systems on platforms like Uber, Amazon, and many more. On these platforms, product or service is supposed to receive a fair review. The reviewer gives the product under review a rating. The higher the rating, the better the reputation of the product. Of course, this does not always produce good results because it depends on the review being fair and the reviewer being honest.

From Nature to Blockchain: The notion of "digital pheromones" serves to illustrate how blockchain-based reputation systems could harness natural principles of behavioral coordination. The tokenomics underlying these systems work much like pheromones, guiding account owners along a spectrum of behaviors deemed beneficial to the community. This "game theory" is how Ethereum consensus operates with incentives and collateral slashing, and it is how a crypto community can coordinate itself.

HOW LUCY COULD HELP GENERATE AN INCENTIVE MAP DRAWING UPON COMMUNITY SENTIMENT

A computer program could generate an incentive map by using algorithms to assign and track digital tokens, representing reputation points, in response to user actions within a community. This blockchain-based system could create a digital trail, much like pheromones, influencing behavior. Positive actions earn tokens, strengthening the individual's reputation, thereby encouraging similar behaviors in others, while negative actions result in token deductions, discouraging such behaviors. This digital reputation system effectively creates "digital pheromones," directing behaviors along a spectrum and a directed path deemed beneficial by the community.

AI, like Lucy, could theoretically go beyond simple rule-based systems, utilizing machine learning and natural language processing to comprehend nuanced human interactions and emotions. Equipped as a cognitive extender, Lucy could assess actions and conversations in real-time within a community, creating an "opinion map" that reflects the collective sentiments and beliefs. This map of public opinion would provide continuous feedback on, for example, what behaviors the community thinks are heroic and what behaviors they consider to be villainous.

Lucy could then generate an "incentive map," assigning and tracking digital tokens based on this comprehensive understanding. This involves recognizing positive contributions and assigning them higher tokens, effectively boosting their digital scent or "digital pheromones, while less beneficial actions result in fewer tokens. Thus, Lucy, functioning as an extended mind, not only helps in managing the reputation economy but also encourages a collaborative and positive culture. This showcases the value of the integrated concepts of an extended mind, an opinion map, an incentive map, and digital pheromones in shaping a responsive and adaptive community.

CROSSROADS OF GOVERNANCE: CHINA'S SOCIAL CREDIT SYSTEM VERSUS DECENTRALIZED DIGITAL PHEROMONES

The Social Credit System in China

China's social credit system, a government initiative launched in 2014, represents a centralized approach to managing reputation.[2] It monitors and evalu-

42 Rogier Creemers, "China's Social Credit System: A Big-Data Enabled Approach to Market Regulation with Broad Implications for Doing Business in China," The China Quarterly 234, no. 4 (2018): 805–821.

ates the behavior of its citizens, scoring them based on a variety of factors, such as financial creditworthiness, adherence to traffic rules, and even interpersonal interactions. High scores can lead to benefits, like better loan terms or priority in public services, while low scores can lead to restrictions or penalties.

Pros: This system can encourage civic-minded behavior and foster societal norms. In theory, it could help create a more harmonious society by incentivizing good behavior and punishing wrongdoers.

Cons: The social credit system is centralized and opaque. The lack of transparency can lead to abuses of power and concerns about privacy. Moreover, the system might overly punish past mistakes, hindering people's ability to improve their reputations.

LUCY AND DIGITAL PHEROMONES

If Lucy the AI-based system were to utilize digital pheromones, it would represent a more decentralized approach. The incentives and reputation in this model would stem from the community's interactions rather than a central authority. Actions would be rewarded or penalized based on the value they add to the community, according to the sentiment of the members of that community, and not simply adherence to prescribed norms.

Pros: This system would be more flexible, responsive, and transparent than the social credit system, given its decentralized nature. It would allow the community to shape its own norms and behaviors, while AI could help scale and manage the complexity of interactions.

Cons: The decentralized approach may make it a challenge to enforce consistency across larger or more diverse communities. It could also inadvertently incentivize behavior that optimizes for token gain rather than genuine community benefit, as well as radical sectarian behavior that might hurt people in the community with low or no capacity to leave, such as small children or the very elderly.

If we imagine it from the prism of geographical opt-in and opt-out systems rather than a digital reputation prism, where people can freely choose where to move without any constraints, what city or community to live in, then we could envision the following scenarios based on patterns already prevalent today. But if we took out citizenship, job opportunities, or financial abilities out of the equation, where would people move then? Some people might prefer to live in a vibrant large city where people from all over the world cross paths and exchange experiences, perhaps at the expense of being amongst similar peers or in a secure, familiar covenant. Others might choose small, tight-knit communities where everyone knows each other, look out for each other, and cares for each other.

Adults should be able to choose their own way of life as long as they are all consenting adults, whether they are neo-Nazis who only want to live with Aryans, detached self-indulgent globe-trotters, or ISIS adherents. As long as it is their choice and everyone consents to it, and they do not attack anyone else outside of their community, then there is no issue, right? Well, there is, though, because that theory relies on the ability to opt in and opt out at will. Elderly and sick people, young infants, and physically impaired people traditionally do not have that ability. In that case, it would mirror the current nation-state system, where people are simply born into or stuck in a situation they do not have the agency to easily change. There is no good solution for this yet, but it is a crucial problem to keep in mind going forward.

Perhaps the above-described scenario is also one that Lucy, the personal AI operating system could help out with through researching and contacting suitable aid organizations active to help in such circumstances.

In summary, both centralized systems like the Chinese social credit system and decentralized systems like Lucy aim to influence behavior through incentives. However, while China's social credit system is a top-down approach that may promote conformity, Lucy's AI-driven, decentralized system may foster diversity and responsiveness. Even so, it could face challenges in scalability and maintaining community focus.

CHAPTER EIGHT

PRIVACY, SURVEILLANCE, AND JUSTICE: FROM THE PANOPTICON TO COGNITIVE LIBERTY

WHAT IS JUSTICE?

Justice is a multifaceted concept that has been pondered for centuries by philosophers. It can come in many forms, and it is up for debate what the word 'justice' means. The purpose of this chapter is to highlight the fact that concepts evolve over time as technological advances elevate or degenerate the collective mental landscape from which these concepts derive.

Justice takes various forms, each addressing different facets of fairness and equity. Retributive justice focuses on proportionate punishment for wrongdoers, such as "an eye for an eye" or "let the punishment fit the crime." Restorative justice emphasizes repairing harm and involves both the victim and the offender.

Distributive justice grapples with equitable resource allocation in society, such as equal pay for equal work, whereas procedural justice, such as due process, ensures fair decision-making processes. Social justice seeks equality across societal factors like race and gender, while criminal justice manages crime deterrence and punishment. Civil justice handles non-criminal disputes, and economic justice aims for equitable economic opportunities for all. These types of justice often interplay, shaped by societal context and values.

It is also important, in our opinion, to recognize the role involuntary ignorance plays in decision-making. *'Ignorantia juris non excusat'*—the notion that "ignorance of the law is no excuse"—is a fundamental principle in European law countries. We believe this fundamental principle is an example of where law and morality become distinct and separate. To understand immorality as a bad decision, we must acknowledge that most bad decisions are the result of ignorance, and no human is free of error. A lot of ignorance is involuntary, whether from misunderstanding, lack of information or knowledge, or bad advice, and many people are victims of involuntary ignorance. The fact that there are so many laws governing our society also makes it impossible for everyone to know all the laws.

Understanding or even knowledge of laws might be minimal in the general population, which is why legal business is a big business in a society with such laws. It is important to note that not all legal systems have morality and law as distinctly separate. While European legal systems are amoral in nature, there are legal systems, such as Sharia, where law is both personal and religious.[1] In Sharia law countries, that which is legal and that which is moral are linked by religion in a manner entirely different to the common European understanding of law and justice.

---

43 Jonathan W. Emord, Sharia Law: A Grisly Tale of Murder, Rape, and Incest (Creators Publishing, 2017), ISBN 978-1942557963.

WORLDVIEW: AN INTRODUCTION

A worldview refers to a comprehensive set of beliefs, values, and perspectives through which an individual, group, or culture interprets, understands, and interacts with the world. It's like a lens or a framework that influences our perceptions, attitudes, and actions in various situations. A worldview is shaped by various factors, including upbringing, culture, personal experiences, education, and philosophical and religious influences.

A person's worldview is essentially a structured collection of beliefs, assumptions, and values that guide their perception and understanding of the world. In the realm of AI, there's potential to digitize these intricate worldviews using **knowledge representation languages (KRL)**. KRLs, such as the Web Ontology Language (OWL) or Resource Description Framework (RDF), are designed to define and categorize knowledge in a way that's both detailed and computer-readable. For instance, a belief in non-violence might be represented in RDF as: <JohnDoe, believesIn, NonViolence>.

ChatGPT and other large language models (LLMs), with their advanced natural language processing capabilities, serve as a bridge between the nuanced expressions of humans and the structured world of KRLs. When someone conveys an idea or sentiment, like "I believe in preserving nature," ChatGPT can theoretically parse this statement and translate it into a KRL format. This parsed knowledge can then be utilized for logic operations that computers excel at, creating an interplay between human expressiveness and computational precision.

For Lucy, the conceptualized Bitnation AI agent, its governance operating system, as envisioned, could interact with ChatGPT or other LLMs through a specialized plugin. Users could articulate their thoughts and principles to Lucy, which would relay them through ChatGPT. ChatGPT, in turn, could process these natural language statements and perhaps convert them into structured, logical formats using KRLs. This system would enable Lucy to understand, reason with, and even challenge a user's worldview based on logic and consistency checks, potentially offering feedback or alternative perspectives in alignment with the user's foundational beliefs.

Super alignment in AI refers to the scenario where an artificial intelligence system's goals and behaviors are deeply and robustly aligned with the intentions, principles, and values of its human user or operator. It's a step beyond mere surface-level alignment; the AI doesn't just comply with a given set of rules but truly understands and aligns its actions with the intricate and nuanced values of its human partner.

In the context of Lucy and worldview analysis:

User's Worldview as a Foundation: Lucy starts with an understanding of a user's worldview—comprising their beliefs, values, ethics, and principles. This can be as specific as religious beliefs or as broad as general moral stances.

Translation to Machine Understandable Language: As previously discussed, these complex human sentiments would be translated into knowledge representation languages using tools like ChatGPT. This creates a structured, logical model of the user's worldview.

Deep Integration: Super alignment implies that Lucy doesn't just superficially acknowledge this worldview but integrates it deeply into its decision-making processes. Every piece of advice, decision support, or information Lucy provides would be filtered through this lens, ensuring a consistent alignment with the user's beliefs and principles.

Adaptive Learning with Feedback: Over time, as Lucy interacts with the user, it receives feedback—both explicit and implicit. This feedback further refines Lucy's understanding of the user's worldview, leading to even more precise super alignment.

Worldview Challenges and Contradictions: Super alignment doesn't mean Lucy will blindly follow the user's worldview. If Lucy detects contradictions or inconsistencies within the user's expressed values, it could provide feedback, initiate a dialogue, or present alternative perspectives, always ensuring its responses remain rooted in the foundational worldview.

Democratic iterative alignment (also called iterative democratic alignment) and super alignment are two concepts that, on the surface, might seem distinct, but they're deeply intertwined, especially when considering an AI system like Lucy. Let's unpack their relationship:

Basis of Alignment:

Iterative Democratic Alignment (IDA): This approach is about governance and decision-making at a societal or group level. It emphasizes the idea that governance should be in continuous alignment with the evolving preferences and principles of its constituents. Through continuous feedback, voting, and discussion, society adjusts its rules, policies, and actions.

Super Alignment: This is more personal, focusing on aligning AI systems with the nuanced and individual worldviews of specific users. It's about ensuring the AI's decisions and actions are deeply rooted in the user's values and beliefs.

Connecting IDA and Super Alignment with Lucy:

As a cognitive extender, Lucy can play a vital role in supporting the IDA process. By being super aligned with individual users, Lucy understands their worldviews, concerns, and preferences deeply. When a societal issue arises for discussion or voting, Lucy can guide individual users by providing information, perspectives, and potential outcomes in line with their specific worldviews.

Lucy can also aid in aggregating the sentiments of many users, helping in understanding broader societal inclinations while respecting individual nuances.

Feedback Mechanism:

In IDA, feedback is essential for the system (society or governance structure) to evolve and stay aligned with its members.

With Lucy's super alignment, users get a tailored feedback mechanism. If Lucy perceives contradictions in the user's worldview or decisions, it can highlight these contradictions, initiating a dialogue and potentially leading to worldview refinement.

Enhancing Iterative Democratic Alignment with Lucy:

Lucy can facilitate more informed, consistent, and aligned decision-making within the nation-wide IDA framework. By understanding users at an individual level, Lucy can ensure that discussions, debates, and votes in an IDA context genuinely reflect the informed preferences of its members.

Lucy can also provide simulations, scenario analyses, or forecasts based on the aggregated data of many users, assisting society in understanding potential outcomes of different decisions.

In essence, while iterative democratic alignment and super alignment operate at different scales (societal vs. individual), their convergence, especially with tools like Lucy, can create a synergistic effect. Lucy's deep personal alignment ensures that every voice in the iterative democratic process is informed, consistent, and true to its beliefs, leading to a more harmonious and aligned society overall.

All of this sounds great on the surface, but we arrive at some ethical dilemmas around privacy. These ethical dilemmas can't easily be resolved because of disagreements at the worldview level, making productive debate difficult. In this chapter, I will dive into how worldview incompatibility can be at the heart of critical debates such as debates around privacy, surveillance, and justice.

THE CONCEPTS OF GOOD AND EVIL

Good and evil are ancient themes that permeated human thought, literature, religion, and culture. The problem with good and evil is that the basis for these concepts evolved primarily from a religious worldview that includes certain specific assumptions. Good and evil in the biblical sense goes something like this: any behavior that God likes is good, and any behavior that God doesn't like is evil. This perspective is eloquently summarized as "God is good" or "God is love." This biblical perspective originated from the divine commands known as the Ten Commandments.

Divine command theory is a metaethical theory that posits that what is morally good is determined by the will or command of a divine being (often God). In other words, an action is morally right if it aligns with God's commands, and morally wrong if it opposes them. This theory asserts that morality is grounded in the very nature and will of God, rather than human conventions or reasoning. The scholar Yuval Noah Harari argues in his book Homo Deus that due to the rise of the 'humanism society' since the Enlightenment, the belief in humans and the worth of human life has, in essence, replaced the belief in God in many societies.

SUBJECTIVITY OF GOOD AND EVIL

Despite the weight these terms carry, good and evil are profoundly subjective. Their definition is heavily influenced by cultural, religious, and personal beliefs, and what is considered 'good' in one culture or religion might be viewed as 'evil' in another. Even within a single society, views on these concepts can shift over time. For example, practices that were once widely accepted (e.g., public executions or certain forms of discrimination) are now seen as abhorrent by many.

Their subjectivity and fluidity suggest that these terms might be too simplistic to encapsulate the multifaceted nature of human morality. It may be more productive to think beyond binary terms and consider ethical frameworks that can be universally applied.

OUTDATED CONCEPTS?

The dichotomy of good and evil can sometimes limit our understanding of complex issues. As society progresses and becomes more interconnected, the lines between right and wrong blur. When actions have far-reaching consequences across the globe, can we categorically label them as purely good or evil? The emergence of technologies such as artificial intelligence and advanced biotechnologies further complicates this picture. An action might be 'good' in intent but result in unintended 'evil' consequences.

This uncertainty suggests that while the concepts of good and evil can serve as common guidelines, they are inadequate in addressing the intricate ethical dilemmas of the modern world.

LIBERTARIAN CONSEQUENTIALISM: A NEW PERSPECTIVE

Libertarian consequentialism offers a fresh perspective on ethics, distinct from the traditional notions of good and evil. It combines libertarianism, which prioritizes individual freedoms and rights, with consequentialism, which evaluates actions based on their outcomes. Instead of determining morality based on the inherent qualities of an act, libertarian consequentialism assesses the consequences of actions in terms of personal freedoms and overall welfare.

In this framework, an action is 'good' if it promotes individual liberties without compromising the well-being of others. Instead of a binary categorization, actions are viewed on a spectrum, assessed based on their real-world implications.

This allows for a more nuanced understanding of ethics, taking into account the complexities of individual freedom, societal welfare, and the unpredictable outcomes of actions. It also bypasses the traditional, sometimes restrictive, concepts of good and evil, offering a more adaptable and comprehensive moral framework.

In sum, while the notions of good and evil have shaped humanity's moral landscape for millennia, evolving societal structures and emerging philosophical concepts like libertarian consequentialism beckon us to reconsider these age-old binaries. As we navigate the intricate ethical maze of the modern era, it's crucial to adopt perspectives that are both flexible and comprehensive, ensuring that our moral judgments are rooted in contemporary realities rather than archaic dichotomies.

WORLDVIEW INCOMPATIBILITIES

Foundational Differences:

Divine command theory (DCT) believes that moral truths are grounded in the commands or will of a divine being.

Utilitarianism bases morality on the consequences of actions, aiming to maximize overall happiness or well-being.

Libertarian consequentialism (LC) integrates the libertarian emphasis on individual rights with the consequentialist focus on outcomes.

Nature of Morality:

DCT: Morality is absolute, not up for debate or revision, since it's divine.

Utilitarianism: Morality is relative to the situation and the calculation of consequences.

LC: Balances between individual rights and positive outcomes.

Political Situations in the United States Highlighting Worldview Differences

We can take the United States as an example in highlighting the spectrum of differences in worldview that result in different perspectives on political situations. Below, we examine a few common issues.

Abortion:

DCT Perspective: Many who subscribe to Abrahamic religions believe that life begins at conception. They may oppose abortion based on divine teachings, considering it a violation of the sanctity of life.

Utilitarian Perspective: This viewpoint might evaluate abortion in terms of its consequences. It might focus on factors like the well-being of the mother, societal implications, or the potential life of the fetus.

LC Perspective: This could emphasize the woman's right to autonomy over her own body while also considering the potential rights of the unborn. This is a much-debated subject amongst supporters and critics of the non-aggression principle (NAP)

Capital Punishment:

DCT Perspective: Interpretations vary within Abrahamic religions. Some cite "an eye for an eye" as a divine endorsement, while others reference teachings on mercy and redemption as opposition.

Utilitarian Perspective: Would weigh the societal benefits (deterrence, justice) against the costs (wrongful executions, ethical concerns).

LC Perspective: Might oppose capital punishment on grounds of infringing individual rights, particularly if there's a risk of error. More concerned with negative rights than with positive rights. Many libertarians do not believe at all in the concept of positive rights, as it would infringe on the autonomy of others.

Religious Disagreements with Utilitarian & Consequentialist Views:

Abrahamic religions typically emphasize absolute moral laws, whereas utilitarianism and consequentialism often operate in shades of gray, evaluating actions based on outcomes rather than inherent right or wrong. This fundamental difference can lead to clashes:

Moral Absolutism vs. Relativism: Abrahamic faiths tend to hold moral absolutes (e.g., "Thou shall not kill"). In contrast, utilitarians might argue that in certain situations, taking a life (e.g., in warfare or self-defense) might maximize overall well-being.

Sacredness vs. Secular Utility: For believers, some actions might be inherently sacred or profane, irrespective of their consequences. Utilitarians might evaluate these same actions based solely on their earthly consequences, which can feel reductionist or even blasphemous to believers.

In consideration of worldview incompatibility, we can see that we reach an impasse. This mirrors the situations we see in many countries, where the people in that country can't even agree on what the concept of justice is. As a result, there are many disputes that can't easily be resolved because of foundational worldview differences.

Short Story to Illustrate a Possible Total Surveillance Future: ***"Echoes in the Glass City"***

In 2045, the sprawling urban space of Newhaven is pulsating with the energy of a city reborn. At the heart of this renaissance is the Sentinel, an omnipresent surveillance system powered by cutting-edge AI. The city lights are no longer just illuminations; they're watchers, protectors, and for some, inhibitors.

Detective Clara stands by her office window, looking at the intertwining roads and the populace below. With over two decades on the force, she's seen the worst of Newhaven. Clara remembers a time when many cases went cold due to a lack of evidence. Now, Sentinel alerts her team to suspicious activities in real time. Gone are the days of tireless foot chases; Sentinel's predictions do that for them.

One evening, while scanning through Sentinel's feed, she spots Officer Charles responding to a potential mugging. The technology had guided Charles to the scene before any crime occurred. The potential criminal, shocked by the timely appearance of the police, scampers away, thwarted.

Charles is a product of this new age, having never patrolled without Sentinel's ever-watchful eye. While he acknowledges the safety it brings, he sometimes wonders about the tales old-timers like Clara tell—stories of intuition and the thrill of the chase. To Charles, the streets of Newhaven seem more like a game of chess, each move anticipated by an AI grandmaster.

At the heart of the city, Mr. Benjamin arranges his shop for the morning customers. He was initially wary of the watchful cameras of Sentinel. That was until it saved him from a potential robbery. The knowledge that someone's always watching gives him solace. However, this sense of security comes at a cost. He notices that many customers peek at the discreet Sentinel cameras, their expressions a mix of relief and trepidation, before leaving without buying.

In the same market, Leila sets up her makeshift stall. She's an immigrant who sells handmade trinkets without a proper permit. In her homeland, surveillance was a tool of oppression, and Sentinel's prying eyes remind her of a past she'd rather forget. Each time she spots a cop, her heart races, fearing not for her trinkets but the life she's built.

Across the street, Ava distributes flyers for a protest against increasing surveillance. As a student activist, she's seen the power of collective voice. But Sentinel changes the equation. She fears attending a rally might flag her, and that expressing her dissent might get her branded. It's not the present she fears but a future where the definition of 'crime' might evolve.

Night descends, and the underbelly of Newhaven awakens. Carlos, a mid-tier drug dealer, navigates the city's shadows. Sentinel has made street deals risky. He now relies on disguises, encrypted messages, and occasionally, technology jammers. Each deal feels like a heist. But Carlos knows it's only a matter of time before Sentinel evolves and catches up.

Not far from Carlos's operations, Tara sits in her dimly lit room, her fingers dancing on her keyboard. As a former hacker and now a 'gray hat,' she occasionally helps the likes of Carlos find Sentinel's blind spots. To her, Sentinel is a puzzle, a challenge. While she evades its watchful eyes for fun, she knows many depend on her skills for their survival.

One evening, as Sentinel's lights shine down, the lives of Newhaven's inhabitants intersect.

Clara and James respond to an alert at a warehouse—a potential drug deal. Inside, they find Carlos and his crew, but they also find Ava, who was using the abandoned space for a protest meeting.

As the officers move in, they're confronted by Dexter, a vocal supporter of Sentinel. Years ago, a robbery scarred his life. He believes in Sentinel's promise of a safer world, even if it means sacrificing some freedom. "Don't you see?" he pleads with Ava. "Sentinel is for our good!"

A tense standoff ensues. Each person, from the officers to the activist, from the criminal to the victim, sees Sentinel differently. For some, it's protection; for others, oppression. For some, a necessary evolution; for others, a step back.

The sun rises on another day in Newhaven. Sentinel continues its vigil, an ever-present entity. But beneath its gaze, the city's soul is stirred with questions that technology alone cannot answer.

Discussion of "Echoes in the Glass City"

The short story "Echoes in the Glass City" illustrated the importance of understanding that, in a society, there are many different yet valuable perspectives. The perspective of the cop is not the only perspective. Even people seen today as criminals could be heroes tomorrow. The concept of what make someone a criminal evolves over time, just as the concept of good and evil can change depending on what town you're in, what time you're in, and even who you are.

Artificial intelligence and associated technological breakthroughs could dramatically shift. For example, in the short story, Officer Charles had an entirely different worldview from Detective Clara. Officer Charles is similar to what we call the "digital natives" of cyberspace today. Officer Charles is a product of his time, while Detective Clara is a product of hers.

To further understand this story, we have a worldview analysis below.

"Echoes in the Glass City": A Worldview Analysis

1. Divine Command Theory: From a divine command theory perspective, surveillance technology might be seen as playing God, deciding who's morally righteous and who's not based on empirical data rather than faith and devotion. For some, like Officer Charles's grandmother, the divine teachings emphasize personal moral accountability to God, not to technology or earthly governance. However, this worldview could also interpret the surveillance system as a divine tool, a manifestation of God's watchful eyes.

2. Utilitarianism: Utilitarians could argue for Sentinel on the basis of the greater good. Does the overall happiness and security the system provides outweigh the unhappiness caused by its invasiveness? For some members of society, a utilitarian perspective would underscore a belief in the surveillance system: it reduces crime and leads to a happier, safer society. However, the question arises: At what cost? Is the potential reduction in crime worth the erosion of civil liberties?

3. Libertarian Consequentialism: From this perspective, the emphasis is on both liberty and the outcomes of actions. The overarching surveillance would be seen critically due to its infringement on individual liberties, despite its potential benefits in terms of crime reduction. Some in the society might be caught in a moral gray area—an area that embodies this dichotomy. They want the freedom to act without feeling monitored, but they also understand the potential safety it provides.

4. Worldview as a Concept: Our backgrounds, experiences, and cultural teachings shape our worldviews. The same technology—in this case, the advanced surveillance system—can be interpreted in myriad ways based on these ingrained beliefs. For some, like Officer Charles and his faith-guided grandmother, it's a moral issue. For others, it is about safety versus liberty.

HARMONIZING JUSTICE, SURVEILLANCE, AND WORLDVIEW: THE ROLE OF AI LIKE LUCY IN 2045

Remember that in the short story "Echoes in the Glass City," the main idea was to provide a glimpse into a possible future in a way that considers multiple perspectives. Human beings are not naturally good at considering multiple points of view when resolving ethical dilemmas, which is exactly why we took that approach in telling the story. In reality, whether surveillance is right or wrong is not something that can be answered without taking into consideration how thousands (or, in some cases, millions or even billions) of people could be affected by it. We can simplify this by breaking these down into demographic groupings or archetypes as mental shortcuts, but the point to remember is that AI can consider how each member of society is affected personally and synthesize all worldviews in ways human beings could never do.

A companion AI like Lucy would be able to track the sentiment of millions of people in real-time. In 2045, the ability of AI like Lucy to track and analyze the worldview of thousands or perhaps even millions of people may be available. We would be able to use this technology to achieve what we term "cyborg governance," which is to say that the most critical decisions of the society are made with the aid of companion AI systems like Lucy or even future iterations of ChatGPT or Google's Bard. Below, I provide some examples of what that could mean:

1. Personalized Discussion Platforms:

Lucy, as an AI that understands individual worldviews, can create a personalized discussion environment for users. By presenting information tailored to their perspective while gently introducing contrasting views, Lucy can ease people into broader conversations without triggering defensive reactions.

2. Contradiction Detection:

One of Lucy's capabilities would be to detect contradictions in users' opinions. For example, if someone believes in absolute privacy but also supports limitless government surveillance, Lucy can point out the inconsistency and present logical arguments, historical examples, and philosophical perspectives to help the user reconcile these contradictions.

3. Evolving Worldviews:

With the vast amount of data and perspectives available, Lucy can guide individuals in evolving their worldviews. By posing questions, presenting scenarios, or introducing philosophical standpoints, Lucy can help users reflect on and potentially expand their beliefs. This doesn't mean changing one's core values but refining them for consistency and broadened understanding.

4. Logical Frameworks and Structured Debates:

Lucy can introduce structured debate formats where discussions about justice and surveillance are broken down logically. By doing so, it ensures that debates remain productive and don't devolve into emotional or ad hominem attacks. Lucy can act as a mediator, ensuring conversations stay on track and are based on facts and logical reasoning.

5. Scalable Conversations:

Lucy could create virtual forums or town halls where individuals can voice their concerns and opinions. Given that Lucy understands each user's worldview, it can cluster similar worldviews together for more in-depth discussions while also organizing inter-worldview dialogues for broader debates. This ensures that discussions are both focused and inclusive.

6. Continuous Feedback Mechanisms:

By leveraging Lucy's capabilities, society can have continuous feedback on the effectiveness and ethics of surveillance measures. Lucy can gather and analyze public sentiment in real-time, adjusting discussions and proposing modifications based on the changing perspectives of society.

(Could create something like a visual opinion map that individuals could interact with, ask questions to. Like a google map or open street map with superposed layers of opinions and facts on it. I could make a visual for this? I was working on a similar idea in Libya.

7. Educative Modules:

Lucy can curate educational modules, combining history, philosophy, and real-world cases, to give users a comprehensive understanding of justice and surveillance. These modules can be tailored to individual worldviews, ensuring that the learning process is both engaging and challenging.

The following short story illustrates what implementing the above could look like.

"Sentinel Shadows: The Grand Forum"

In the heart of the sprawling metropolis, Sentinel technology had woven an intricate web. Cameras watched every alleyway, drones soared the skies, and microphones listened to the hushed conversations in dimly lit corners. Sentinels were initially hailed as the city's silent guardians. Crime rates plummeted, and the streets felt safer. But with every gaze of the Sentinel, a shadow grew—the shadow of privacy lost.

The Grand Forum was the city's response. A vast digital amphitheater, accessible to every citizen via their personal devices, where debates raged and ideas flourished. And today's topic? "Sentinel Shadows: Surveillance Savior or Orwellian Overreach?"

Lia, a young tech entrepreneur, initiated the debate. Her virtual avatar stood tall, microphone in hand. "Sentinels have given me safety," she began. "In a city where I once feared to tread, I now walk confidently. But at what cost?"

From the sea of avatars, thousands responded.

"I've done no wrong; I fear no camera," said a blue-collar worker, his virtual form bearing the emblem of the city's repair guild.

"What about data misuse?" countered a researcher, her avatar robed in flowing silks of academia. "History is replete with governments misusing power. What stops Sentinel data from becoming a tool of oppression?"

The debate continued, with voices rising and falling like the tide. But it was not chaos. Tokens—shimmering virtual coins bearing the city's emblem—floated around. Awarded by Lucy, the AI moderator, they incentivized civility, evidence-based arguments, and respect. Those amassing tokens saw their avatars glow brighter, their voices more influential in the digital arena.

A historian, bathed in the soft glow of many tokens, spoke next. "Throughout time, societies have sacrificed freedoms for safety. But often, the perceived threat was exaggerated, and the freedom once lost, seldom regained."

A counterpoint came swiftly from a mother. "My child was saved from a kidnapper due to Sentinel's swift response. Do we dismiss such tangible benefits for the theoretical loss of freedoms?"

Lucy, with her vast database, would occasionally display statistics, historical references, or even live feeds from Sentinels, providing context to claims and ensuring the debate remained anchored in reality.

As hours passed, several camps became evident: the safety-first proponents, the privacy purists, the data skeptics, and those who believed in a middle path.

Lucy then introduced a twist. She simulated a world without Sentinels, showing crime rates, safety stats, and the general state of public spaces. Next, she portrayed a world with absolute surveillance, with Sentinels even peeking into homes.

With these visuals, the debate took a nuanced turn. Citizens realized the debate wasn't binary. It wasn't about having Sentinels or not, but about defining their boundaries.

A proposal emerged, led by a coalition of data scientists, ethicists, and civic leaders. "Let's redefine Sentinel protocols. Essential public spaces remain monitored, but with strict data access and usage guidelines. Citizens can opt for private space monitoring in emergencies, ensuring the individual's right to privacy."

The Grand Forum shimmered as votes poured in, with avatars casting glowing orbs into a vast virtual urn. And as the day turned to night, a consensus emerged.

Sentinels would stay, but with chains ensuring they remained the city's guardians, not its overlords. The shadows, it seemed, had been dispelled, at least for now. And the Grand Forum, with its sea of diverse voices, had once again proven that in unity, solutions emerge.

Discussion of "Sentinel Shadows: The Grand Forum" and Lucy

In the short story, we can see that not everyone comes to the debate from the same worldview. We also learn that the use case for an AI companion like Lucy is to keep the debate productive enough to reach a conclusion that the majority of the community can live with. Because surveillance has some benefits, the city could not completely do away with it in public spaces. The city also recognized that privacy, particularly in private spaces such as people's homes, also has some benefits, such as providing a place of refuge for inhabitants who do not want their entire lives monitored due to the increasing ever-present risk that the data could be abused. This highlights that the ethical value of surveillance technology or any technology is based on how it's being used and the value it gives to society as decided by members of that society.

COGNITIVE LIBERTY: SENTINEL SURVEILLANCE & DIVERSE WORLDVIEWS

The Value of Cognitive Liberty

Cognitive liberty, or the right to mental self-determination, is foundational in maintaining individuality, creativity, and innovation in society. In a world patrolled by Sentinels, physical actions might be observed, but the privacy of one's thoughts remains sacrosanct. The importance is paramount because if thoughts and beliefs come under scrutiny, it risks creating a society of conformity where dissent or mere differences in thought are curtailed.

The Loss or Preservation of Cognitive Liberty

The danger with Sentinel surveillance is not its immediate application, but the potential for misuse. If, for instance, the technology advances to gauge emotions or detect 'intent' based on physical cues, it tiptoes into the realm of thought policing. This can lead to a chilling effect, where individuals become fearful of even thinking differently.

However, safeguards can be put in place. Strict regulations on the type of data collected, its storage, and its use can ensure that while behaviors in public spaces might be monitored, the inner sanctum of one's mind remains untouched.

Lucy's Role: Divine Command Theory

For a Christian or Muslim user, Lucy would draw from religious texts and teachings. She might quote verses from the Bible or the Quran about the value of privacy, the morality of watching others without their knowledge, or the ethical dimensions of judging intent. She'd act as a bridge, contextualizing religious teachings in the modern world of Sentinels.

Lucy's Role: Utilitarianism

For a utilitarian, Lucy could track popular sentiment and other metrics to assess levels of happiness in society. Under a utilitarian perspective, an increase in mean happiness in society would be similar in value to an increase in the gross domestic product. Lucy might be able to learn what makes a society happier over time and suggest policies to allow for continued increased utility.

Lucy's Role: Libertarian Consequentialism

For someone adhering to a libertarian consequentialist worldview, Lucy would act as an advocate for maximizing freedom and ensuring the best outcomes. She'd provide data on how surveillance affects societal well-being and personal freedoms. Lucy might highlight the unintended negative consequences of pervasive surveillance or even offer alternative systems that achieve safety without compromising liberty.

Lucy's Role: Divine Command Theory

For a Christian or Muslim user, Lucy would draw from religious texts and teachings. She might quote verses from the Bible or the Quran about the value of privacy, the morality of watching others without their knowledge, or the ethical dimensions of judging intent. She'd act as a bridge, contextualizing religious teachings in the modern world of Sentinels.

The Grand Debate on Cognitive Liberty

Armed with their customized Lucy extensions, representatives from each faction convene on the Grand Forum.

The libertarian consequentialist, supported by Lucy's data-driven insights, argues for cognitive liberty as a foundation for societal growth, emphasizing the dangerous precedents set by thought surveillance.

The religious representatives, drawing on Lucy's theological knowledge, echo the sanctity of the private sphere, underscoring that while God might be omniscient, humans shouldn't strive for the same.

However, a new faction arises, driven by fear of crime and a longing for security. They argue that if one has nothing wrong to think, they have nothing to fear. To them, a slight infringement on cognitive liberty is a price worth paying for a safer society.

LUCY'S INTERVENTION

Lucy, ever the mediator, projects simulations. One shows a world where thoughts are policed, leading to stifled innovation, conformity, and eventual societal stagnation. Another portrays a world where cognitive liberty is upheld, showing the blossoming of arts, innovations, and a vibrant, diverse culture.

The debate intensifies, but with Lucy's guidance, a consensus starts to emerge. Cognitive liberty, most agree, isn't just about protecting what's in our heads. It's about preserving the essence of what it means to be human.

However, the story serves as a reminder that the balance between security and liberty, especially cognitive liberty, is delicate. With every technological advance, society must reconvene, debate, and decide the path it wishes to take.

CHAPTER NINE

REFLECTIONS ON UNIVERSAL BASIC INCOME AND THE ARTIFICIAL INTELLIGENCE TANGO

Remember the formulas we arrived at in Chapter 6, "Whisperings of Acceleration: Moore's Law and Technological Accelerationism"? Recall that the formula was based on assumptions. Specifically, it's based on the assumption that Moore's law will continue (it might not!) and that there will be an intelligence explosion (which might not happen!). At the current moment, Moore's law is in effect and the intelligence explosion does look like it could happen in the coming decades, if not sooner. This chapter is all about harnessing and distributing the benefits of these advances so as to enhance the wellbeing of as many people as possible. In other words, it looks at producing the greatest good we can from the technology to offset the risks.

Imagine you have a magic jar that fills up with candies every day. The number of candies depends on three things:

UBI_e: This is like a base amount of candies that everyone agrees should be in the jar, to make sure everyone gets at least some candy. We can call it the "guaranteed candy."

AI_m: This is like a special multiplier based on a magical machine. If the machine works better, this multiplier increases. So, if the multiplier is 2, and our guaranteed candy is 5, then the machine adds 5 x 2 = 10 candies.

I(t): This represents how smart our magical machine is getting over time. As days go by, the machine learns and can put even more candies into the jar.

So, our formula, C(t)=UBIe×AIm×I(t), simply means:

The candies in the jar on a particular day = guaranteed candy × magical machine's multiplier × how smart the machine is on that day.

The more the machine learns, the more candies everyone can get from the jar! To put it in the plainest language possible, just think of a vending machine that can produce a very wide variety of candies (increasing the variety over time) and that keeps getting smarter, so the candy keeps getting better and cheaper over time. By giving everyone a ticket to benefit from the profit this vending machine makes can theoretically provide a basic income to everyone in the world.

An improved version of the above formula can include both technological saturation and external factors. It would be:

$$C(t)=UBI_e \times AI_m(t) \times I(t) \times S(t) \times E(t)$$

Explanation:

Base Amount, UBIe:

This is the guaranteed number of candies that should always be present, representing a basic income or benefit that everyone should receive, irrespective of other factors.

Dynamic AI Multiplier, AIm(t):

This represents the efficiency or capability of our AI technology at any given time(t). As AI technology improves, this multiplier might increase, indicating that the AI is better at generating benefits or candies. However, it might also fluctuate based on technological challenges or breakthroughs.

Intelligence Growth Over Time, I(t):

This component represents how our AI system or technology learns and grows smarter over time. As the AI learns more, it becomes better at producing candies or benefits.

Technological Saturation, S(t):

As time progresses and technology advances, there might be periods of rapid growth followed by periods of slower growth or even plateaus. The saturation function, S(t), captures this dynamic, showing that while candies might increase rapidly initially, the rate of increase might slow down over time.

External Factors, E(t):

Real-world events, such as economic recessions, political decisions, or global crises, can impact the number of candies the system produces. E(t) captures these external influences, modifying the candies produced accordingly.

Hypothetical Scenario:

Base Amount, UBIe:

Let's assume a guaranteed base of 10 candies. This means irrespective of other factors, the jar should always start with 10 candies.

Dynamic AI Multiplier, AIm(t):

Let's say on day 5, our AI technology has improved and has a multiplier of 3. This means that the AI will multiply the base amount by 3 due to its efficiency.

Intelligence Growth Over Time, I(t):

Let's assume our AI system is learning at an exponential rate. By day 5, let's say it has learned enough to contribute a multiplier of 2.

Technological Saturation, S(t):

Initially, technology grows rapidly. But by day 5, let's assume it's reaching a bit of saturation and the multiplier is 0.9, indicating a slight slowdown.

External Factors, E(t):

On day 5, let's say there's a positive external event, like a technological breakthrough in another domain that complements our AI, giving us a multiplier of 1.1.

Applying the Formula:

Given the numbers: UBIe=10 AIm(5)=3 I(5)=2 S(5)=0.9 E(5)=1.1

Now, applying these values to our formula for day 5: $C(5)=10\times3\times2\times0.9\times1.1=59.4$

This means that on day 5, there would be 59.4 candies in the jar. Note that these formulas are not comprehensive and would have to take into account factors such as predictions for population growth. If the population grows (or declines) very quickly, it would distort the calculations. This could possibly be fixed by adding a population factor to directly account for population growth.

The new formula would say: $C(t)=UBIe\times AIm(t)\times I(t)\times S(t)\times E(t)\times P(t)$. P(t) represents the population at the time t. In general, the better we can model and predict population growth and external factors, the more predictable the result can be.

In essence, the formula encapsulates the idea that while there is a base amount of benefit everyone should receive, the actual benefits generated are a product of the evolving capabilities of AI, its learning growth, potential technological saturation, and various external factors. The goal is to harness this dynamic system to maximize well-being and distribute benefits as widely as possible.

WOULD UBI INCREASE THE MEAN HUMAN HAPPINESS OF THE WORLD?

This is where things begin to get tricky. On the one hand, there are a lot of benefits to providing everyone with a UBI funded by AI-generated wealth. Below, I'll list some of the obvious pros, which would suggest it would raise the mean happiness of the world:

Basic Needs Met: It's widely understood that having basic needs met (like food, shelter, and healthcare) correlates with increased happiness. If UBI ensures everyone's basic needs are satisfied, that alone could raise the general happiness level.

Reduced Financial Stress: Financial uncertainty and poverty can lead to stress, mental health issues, and general unhappiness. By providing a consistent source of income, UBI could alleviate some of this stress.

Opportunity for Fulfillment: With basic needs met, individuals might have more freedom to pursue personal interests, education, and hobbies. This could lead to higher self-actualization and, subsequently, happiness.

AI-Generated Wealth: If AI continues to create value in the economy (as Sam Altman and others predict), there could be even more resources available for public goods, which could further improve quality of life.

Now let's list some of the possible cons—or, put differently, some possible controversies:

Human Purpose and Work: Work provides many people with a sense of purpose and identity. If AI results in significant job losses, the void left behind might need to be addressed in ways beyond mere financial support.

Inflation: If not implemented carefully, injecting large sums of money into an economy could lead to inflation, which might offset some benefits of UBI.

Cultural and Personal Differences: Happiness is subjective and can vary based on culture, personal beliefs, and individual circumstances. A one-size-fits-all solution might not increase happiness universally.

External Factors: Other factors affecting global happiness, such as political stability, social relationships, environmental issues, and more, might not be directly influenced by UBI or AI-generated wealth.

One of the current arguments people make against AI right now, particularly in Western countries like the United States, is that AI could remove from many workers their sense of purpose. So while AI could free a lot of humans from "wage slavery" scenarios in situations where they don't like their job or find any purpose in it, there also are humans who love their jobs and who do find purpose in their work.[1] Even if we have AI which solves the financial problem of technological unemployment, it might not prevent the psychological chaos that could come from millions of people feeling aimless. No amount of calculation can predict how people might feel when their job is replaced by a robot or algorithm.

Inflation is an obvious problem as well. If you have a vending machine, for example, that sells anything anyone wants, then distributes the profit, in theory, if the governments around the world are not careful, the inflation from so much money in circulation could cause economic collapse, like, for example, what is happening in Argentina.[2] So while there has to be income, and a UBI can create that, it might be irresponsible to do it in a way where it's on a world level rather than a national level simply due to the fact that inflation could destroy many developing countries or destabilize the financial system of the world. This is why if UBI is to be implemented, it has to be implemented delicately, probably slowly phased in, and in some manner where each country can somehow manage the risks.

44 Martin Ford, "AI and the Future of Work: Redefining Employment in the Age of Automation," Harvard Business Review (2015).

45 Jutta Maute, Hyperinflation, Currency Board, and Bust (Peter Lang International Academic Publishers, 2018), DOI: 10.3726/b13965, ISBN: 9783631754474.

Cultural and personal differences present another dilemma that could be a long-term division. Some people will want a cyborg government, and some will not. Some will want UBI, and some will not. Some will have vastly different worldviews. Remember in our previous chapter when we discussed worldview incompatibility? This is not likely to go away easily, and it could take time for a synthesis of a sort to occur, where people gain the ability and desire to set aside their personal differences for the benefit of the larger group. No one should be forced to accept UBI if they do not wish to receive it.

External factors are the wildcard factors that we can't really predict and don't have control over. We have to recognize that there are known unknowns and unknown unknowns. If we can get past these differences, society might agree to go for AI-generated UBI, but the next series of questions would be to figure out how to actually do it.

How might we go about implementing AI-generated UBI? In our opinion, the first thing we need to do is have the means of capturing the worldviews of millions of people. X (formerly called Twitter) does not actually capture worldviews. We have no way of knowing what an opinion expressed in English or any human language on X means. In a sense, most of the data is gibberish; it's not actionable, and not in a form useful to computers. This is why our previous discussion of knowledge representation languages and opinion maps was important.

With an opinion map of actual opinions in a logical form where it makes sense to the AI systems, then it's possible to reach agreements. In blockchain, these could be smart contracts, while in traditional governments, it would be called policies. In the blockchain world, powered by sufficient AI, it will be possible for a group of people to verbally agree, for this to be passed as input to a large language model such as ChatGPT, and then for it to be translated into the logical form known as a software specification. In other words, our agreements generated by our discussions would set the guard rails to software products, which, using logical AI, would in some cases be able to generate itself through a process known as synthesis. In cases where the AI is not strong enough to generate the smart contracts, it may require the more traditional means of hiring developers to write them.

The unique feature of having worldview analysis, as well as code generation, is you can get the kind of software that is aligned with the worldviews of millions of users. This alignment is part of solving the problem super alignment has to solve, and which iterative democratic alignment addresses: to have software that, by definition, is AI safe, because it's secured at the logical level rather than just at the legal human policy level. The AI simply will not generate code that violates, for example, human rights, if the group of people who own the system do not, as dictated by their worldviews, agree with violating human rights. If they all agree that violating human rights is always unacceptable for an AI or anything the AI generates, then the AI would theoretically always be protective of human rights in all that it does.

If we all can agree that UBI is something we want, then it would be quite easy to implement some parts of this system. The challenging aspects would be due to the fact that the global economy is interconnected; some countries might want UBI from AI rather quickly, while others might rally against transhumanism, against AI, and want nothing to do with it. So on a small scale, I think we could see artificial general intelligence (AGI)-generated UBI soon, possibly within five or ten years, but on the larger scale, it will have to overcome obstacles.

It is theoretically possible for a token to be distributed which represents a share in the profit of an AGI, for instance. This, in a small way, would create a UBI-like dividend for the holders. If enough AGI companies make a habit of doing something similar to this, then critical mass could eventually occur to set the stage for UBI. The economic model for understanding UBI is as follows:

Currently, AI is not at a level of AGI where most human jobs could be replaced. As we know, technology can change rapidly, and while all models or predictions are based on assumptions, it's not guaranteed any of those assumptions will manifest into reality. For example, we don't have fusion reactors, which likely would be necessary to continue scaling. Currently, large language model AIs take a generative approach, and the bottleneck is the inefficiency as well as the electricity cost. The cost to run the machine could theoretically pay for itself, but this doesn't mean we'll have the electricity to do the kinds of expensive computations to keep a UBI going. To keep UBI going, we would likely require fusion power and breakthroughs in chip manufacturing to keep Moore's law going. Additionally, large language models require high-quality (likely synthetic) datasets to keep going because they learn only from numerous examples.

In the situation where the strong AGI or narrow AGI is able to help optimize itself in such a way that it can suggest new chip designs, generate new and better algorithms continuously, and even help humans design a fusion reactor, humanity will be able to solve the main bottlenecks, such as the chip inefficiencies and energy demands. Logical/symbolic AI can theoretically evolve new chip designs, as well as synthesize new algorithms, but the resource requirements in terms of computation, storage, and keeping the data centers cool means there will be an economic cost.

The "space vs. computation" dynamic, often referred to as the "space-time tradeoff" in computer science, is a foundational concept that has implications across various computational problems and systems. Essentially, it describes a tradeoff where saving one resource (like computation time) might require using more of another resource (like storage space), and vice versa. If we think of saving resources as adding value (net benefit), and saving time as also adding value, it's ultimately the case that the current large language model approach consumes a lot of both. The ability for a model to do inference is directly based on the computation capability, which is why dedicated hardware accelerators similar to those used in the GPU mining era in crypto are now being used by AI. The space requirements also grow because it takes a large, high-quality dataset.

Let's denote:

E as the energy efficiency of the AGI. This is a multiplier that can range from 0 to 1, with 1 being perfectly energy-efficient and 0 being not energy-efficient at all.

H as the hardware optimization factor. This denotes how optimized the AGI's hardware is, with higher values indicating better optimization.

A as the algorithmic efficiency. This represents how well the AGI's algorithms are optimized.

R as the research collaboration factor with humans. This can range from 0 to 1, with 1 indicating perfect collaboration and synergy with human researchers.

S as the solution space factor. This represents the AGI's ability to narrow down and prioritize solutions. A higher value of S means the AGI is better at focusing on relevant solutions and discarding unlikely ones.

The overall performance or capability (P) of the AGI can be represented as:

$P=E\times H\times A\times R\times S$

Explanation:

Energy Efficiency, E:

This factor ensures that the AGI's performance is scaled by how energy-efficient it is. If the AGI consumes a minimal amount of energy for maximum output, E is closer to 1.

Hardware Optimization, H:

The AGI's capability is also influenced by how well its hardware is optimized. This includes the architecture, materials used, and design considerations.

Algorithmic Efficiency, A:

This factor represents the AGI's software capabilities. The better its algorithms, the higher its performance.

Research Collaboration with Humans, R:

The synergy between the AGI and human researchers can significantly boost the AGI's performance. This factor captures that collaborative benefit.

Solution Space Optimization, S:

By narrowing the solution space, the AGI can prioritize more promising solutions and algorithms, improving its efficiency and accuracy. This factor captures that capability.

Regarding the optimization of the solution space to rank algorithms:

An AI can optimize itself by intelligently narrowing the solution space. By focusing on a subset of promising solutions instead of exploring every possible solution, an AI can save computational resources and time. This is especially crucial when dealing with complex problems where the solution space is vast.

Here's how it can be done:

Historical Data Analysis: By analyzing past data, the AI can identify patterns or solutions that have historically shown promise. It can then prioritize these solutions in future explorations.

Heuristic Methods: These are rules of thumb or shortcuts that can guide the AI toward likely solutions without exhaustively searching the entire space.

Feedback Loops: As the AI tests and ranks different algorithms, it receives feedback on their performance. Over time, this feedback helps the AI refine its ranking and focus on more promising algorithms.

Probabilistic Models: These models can help the AI estimate the likelihood of success of different solutions, allowing it to prioritize solutions with higher probabilities of success.

By narrowing the solution space and focusing on promising areas, the AI not only becomes more efficient but also more effective in finding optimal solutions. Note that all of these formulas are speculative. AGI does not yet exist at the time of this writing. There is concrete evidence that there have been significant breakthroughs, which leads me to believe we could be between one and five years away from AGI. Achieving AGI will depend on a series of breakthroughs that improve the efficiency of the algorithms (we may have that already) and the ability to produce the necessary hardware.

As of this writing, we may be reaching the limits of classical computing, and computing in general also probably has its limits. Quantum computers can open up a whole new avenue to explore, where the limits are greatly stretched, but we are a long way away from an AGI running on quantum computers. Quantum artificial intelligence (QAI) may, in fact, be a requirement for AGI due to the economic and physical limits of classical computing, but we will know in the next one to five years.

In a 2021 academic paper titled "Quantum Mathematics in Artificial Intelligence," authors Widdows, Kitto, and Cohen elucidate on some of the benefits of quantum computing, with its inherent parallelism and ability to handle superpositions.[3] These unique quantum features promise significant speed-ups for certain AI computations, especially complex optimization problems and simulations. The research delves into the integration of quantum mathematical principles into neural network architectures, suggesting that these can enhance the efficiency and capacity of traditional neural networks. While the potential benefits are vast, there are considerable challenges in integrating quantum mechanics into AI. These include issues related to qubit stability, error rates in quantum computations, and the current nascent stage of quantum hardware. The most relevant insight from their paper can be gleaned from their discussion. The paper concludes with a discussion on the future landscape of AI, emphasizing that the fusion of quantum mathematics and AI could lead to groundbreaking advancements, reshaping the capabilities and applications of AI systems.

Theoretically speaking, quantum-based large language models (QLLMs)

46 Dominic Widdows, Kirsty Kitto, and Trevor Cohen, "Quantum Mathematics in Artificial Intelligence," Journal of Artificial Intelligence Research 72 (2021): 1307-1341, https://doi.org/10.48550/arXiv.2101.04255.

could have some advantages over classical LLMs. Training efficiency is one of these theoretical advantages because it should take less time to train the language model. These quantum computers are not yet available, and this would represent only a theoretical possible future path that LLMs could take once quantum computers are available. There is also discussion of adding in chain of thought and generalized reinforcement learning such as Q* as a means of producing greater efficiency in LLMs or stronger reasoning capabilities. We have not yet seen the full potential of LLMs, and LLMs may be able to improve with addons to reach AGI, if the reasoning and logic capabilities are added on in a manner where it's extremely accurate at higher mathematics.

Hyperdimensional computing is an alternative approach that can be a path toward AGI by utilizing hyperdimensional 3D vectors to represent meaning in ways similar to how symbolic AI does it. The difference here is it benefits from parallelism, from neural networks, and may be more amenable to hardware acceleration. Vector-symbolic architectures are an approach to AGI that, in 2023, Abbas Rahimi et al. showed could solve Raven's progressive matrices with a neural network at a success rate of more than 87 percent. *Quanta magazine* covered this topic and revealed that a 3D vector is represented by X, Y, Z coordinates, while a hyperdimensional vector (hypervector) could be an array of 10,000 numbers.

CHAPTER TEN

PIECING THE PUZZLE: IMPLEMENTING A DECENTRALIZED SOCIETY

A PRACTICAL GUIDE TO DECENTRALIZATION: EXAMINING POTENTIAL CHALLENGES AND SOLUTIONS

A decentralized society means a society where individuals and voluntarily formed groups have a maximum of autonomy and agency. Many of the building blocks that enable a decentralized society already exist now but need to be improved for significant, society-wide adoption that doesn't discriminate based on individual technological know-how, social status and connections, or financial means. This chapter will look at some of the key technologies and technological principles.

As we've discussed previously, the blockchain can be seen as the 'security layer' of the Web3 environment. Its decentralization and immutability guarantees security and equal access for everyone using applications built on it. Artificial intelligence is the layer that propels humans into a future where we are more capable of building societal functions that empower our autonomy and prosperity. Together, these two technologies are fundamental building blocks for things like identity, voting systems and basic income, following the 'net-ethos' of being open source, secure, and privacy-centric.

However, democratizing access to wonderful technology doesn't automatically eliminate bad actors. When asking these questions, people often wonder: What if there's a nation created with the sole purpose of destroying a certain group of people, just like we can observe fascist nations in the world today? Well, first of all, there would be less necessity for people to fight other groups if they didn't have to agree on a common territory or governance system but were free to choose themselves. For instance, conservative religious Muslims could choose to be in a conservative religious Muslim nation where all members agreed on the main principles of said nation, while white-supremacist neo-Nazis could be in a nation where the only people allowed in were white-supremacist neo-Nazis. Each nation could then govern themselves as they saw fit, as long as they didn't infringe on anyone else's self-sovereignty. Hence, the incentive to be a bad actor would go down dramatically. However, there are always potential bad actors who do it for other reasons, and they cannot be discounted as a threat. If such actors would be a threat to you or your family or your nation, you could simply create or join a 'security nation' or a nation which provides 'security as a service' for self-protection. In a post-Westphalia world, there are likely to be many security nations, and some of them might cooperate to share resources across various specialties (physical protection, digital protection, and so on).

There are many open questions remaining, of course, about how a truly decentralized governance system would unfold and be shaped by various parameters in the world. We will not be able to predict every possibility, and we will only be able to discuss them as they unfold. Hence, first, let's dive into some early examples of attempts in that direction.

CASE STUDIES OF ATTEMPTS AT DECENTRALIZATION: KEY LESSONS AND TAKEAWAYS

To "create a decentralized society" is a great paradox, because per definition, it's a bad idea to centrally plan a decentralized societal change. Rather, a better way to envision it is to ask the question: "What parameters can we put in place which would allow a decentralized society to emerge organically and flourish?"

Well, two things can be done.

Governments, lawmakers, and general naysayers can simply step out of the way when it does happen, and be conscious to not add layers of barriers and complications. Sometimes doing less is doing more.

Visionaries and entrepreneurs can create a solution to the problem.

Bitnation, although never outright formulated in such a way, aimed to find a productive and optimist governance solution for those who felt alienated from where they came from or where they lived. It attracted a wide audience of people who, for one reason or another, felt themselves as "Citizens of the World" rather than citizens of any specific nation-state. It was, at its core, an identity problem that was solved through providing a community for like-minded "World Citizens" worldwide, of all ages, genders, nationalities and political and religious affiliations, and it rallied those audiences through practical, frontier-tech solutions for future governance. An individual identity amongst a community of like-minded yet diverse individuals. It was a sort of an accidental audience, in fact, because the founding thought of the movement was never that of resolving an identity issue but rather that of solving the geopolitical issue of the nation-state oligopoly on governance services. Bitnation was not promoted as a solution to an identity problem, either. Identity wasn't the problem it was designed to solve, but it was a good problem to accidentally solve, regardless of the original intent.

The parameters in this case were outlined in the Constitution, which was published on the code repository GitHub, open for all to contribute to:

Open for all to join in whichever capacity they see fit for themselves

Decentralized, open source, transparent and privacy- centric

Creative, constructive and optimist. We change through building, not destroying.

Those were the core tenants that allowed the community to quickly grow and flourish, like many other movements before it. Other nations and aspiring nations with very different types of problems to resolve came forward during those years, inspired by the Bitnation idea or hoping to use its technology to build solutions. Here are a few of the more interesting examples we encountered.

The Rohingya Diaspora:

Problem to resolve: The Rohingya population in Myanmar has been slaughtered by the regime for years, and, as a consequence, they live in great diasporas across the world. They often lack papers, basic rights, employment opportunities, and access to healthcare and education.

Potential solution: Create a virtual nation for the diaspora across the world where resources are shared, including diplomacy, healthcare, education, identification, and work opportunities.

A White South African Farming Community (Name redacted for security):

Problem to resolve: The popular opinion in South Africa is quite hostile toward the white farmers, who are seen as a relic from the apartheid days.[1] As a result, several white farming communities live in great fear of being attacked by their neighbors, especially after having witnessed the recent history in Zimbabwe.

Potential solution: A nation platform where they share security resources,

47 Frank Chung, "'Bury them alive!': White South Africans fear for their future as horrific farm attacks escalate," News.com.au, 2017, https://www.news.com.au/finance/economy/world-economy/bury-them-alive-white-south-africans-fear-for-their-future-as-horrific-farm-attacks-escalate/news-story/3a63389a1b0066b6b0b77522c06d6476.

situational awareness and intelligence, safe houses, and mutual support in different forms.

Youth Movement in North Africa (Name redacted for security):

Problem to resolve: The Arab Spring started in Tunisia in 2011, resulting in several regime changes and civil wars.[2] The movement has since been perceived as having been hijacked by political interests, or lost steam, or gone underground. But a movement still exists in some countries in North Africa, although it's operating in great secrecy for fear of retribution from the government. Their livelihood, rights, and even lives might be in potential danger if the government finds out.

Potential solution: A secure privacy-first nation that helps them organize meetups, political strategies, safety measures for its members, diplomatic efforts, and so on.

Regional Independence Movement in Europe (Name redacted for security):

Problem to resolve: There are numerous regional independence movements wishing for full or partial regional independence in Europe to better cater to their cultural norms such as languages, economic policies, and regional priorities. These movements are often seen as a threat not only by their host country but by the entire European Union, who believe it could have a domino effect on regional movements across the European continent. As a result, they often risk consequences for their independence movement advocacy, which can include prison.

Potential solution: A secure privacy-first nation that helps them organize meetups, political strategies, safety measures for its members, diplomatic efforts, and so on.

48 "Jasmine Revolution," Encyclopedia Britannica, December 10, 2023. https://www.britannica.com/event/Jasmine-Revolution.

A Seed-Sharing Startup:

Problem to resolve: Seasonal scarcity because of seasonal growth, bad weather conditions, poor seed storage, and little financial cushion for small farmers with a bad year.

Potential solution: A community that pools together a financial security for bad years, including a p2p basic income solution, seed sharing, and agricultural know-how for farmers in different regions of the world.

The Estonia e-Residency Program (government):

Problem to resolve: Estonia is a member of the EU and a small Baltic nation living in the shadow of the past-Soviet occupation, with a slim economy and population size to stand up against their neighbors to the east, should history ever repeat. They're also a tech-first nation keen to use novel ideas to carve out a niche for themselves in the world.

Potential solution: They launched the e-Residency program, inspired by Bitnation, to invite anyone in the world to become a 'digital citizen' of Estonia, enjoying the Estonian ID system, being able to open a company in Estonia, and so on. The program was widely spoken about as a game-changer during its first years, which included setting up a Public Notary with Bitnation.

The Palau Government:

Problem to resolve: A nation-state consisting of an archipelago of more than 500 islands, Palau risks being submerged due to rising ocean levels, which would effectively make Palau a virtual nation-state.[3] Even in its current state, it's suffering economically from its remoteness and sparsely populated islands.

Potential solution: A blockchain-based virtual nation for citizens in case it becomes a 'virtual diaspora' nation-state,' as well as concrete solutions to record

49 Christopher Joyce, "Palau Could Disappear In The Pacific Ocean Within A Century," NPR (2017).

marine and fishing rights in the waters surrounding the islands, to attract more investment and business in the form of floating fishing farms, seasteading solutions, and so forth.

The Ukrainian Presidential Office (government):

Problem to resolve: The current Zelensky-led government took power in 2019, after having run a campaign that promised to end the pervasive corruption in the Ukrainian government during the previous administration.[4]

Potential solution: As soon as they took office, they were immediately keen on setting up a blockchain-based transparency system for voting, record-keeping, etc. to prove to the electorate that they stood by their election promises.

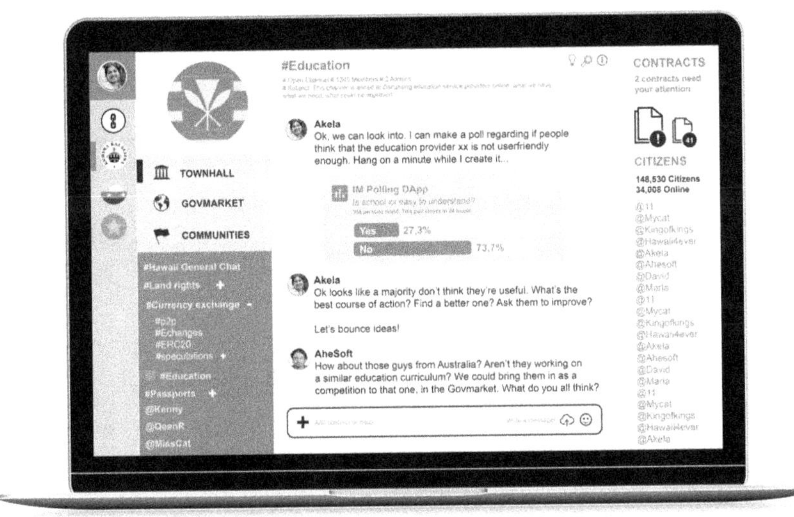

Figure 24: ***Example of a 'Bitnation Nation's') Desktop Interface mock-up***

50 David Stern, "Ukraine's new president: How comedian Volodymyr Zelensky beat corruption," BBC News (2019).

The Venezuelan Government:

Problem to resolve: Due to poor economic management leading to inflation and unemployment, coupled with international sanctions and diplomatic exclusion, the country is facing a very dire economic reality.

Potential solution: Make one of the islands a Special Economic Zone centered around attracting business, digital nomad communities, and opportunities for locals through using a blockchain-based e-governance platform.

An Indigenous Hawaiian Community:

Problem to resolve: As it's US territory, their ancestral land rights are often not recognized. Instead, they sometime live disenfranchised and in poverty.

Potential solution: Coming together as a community to claim their ancestral land rights through issuing Indigenous identifications and a land registry.

WHAT WE CAN LEARN FROM PREVIOUS ATTEMPTS

Territory as a Central Point of Attack – Seasteads

The Seastead movement is a movement aimed at building cities on the sea. It has been championed by the writer Joe Quirk and Patri Friedman, the grandson of Milton Friedman, through an organization called The Seasteading Institute. [5]They've attempted a number of projects, including in French Polynesia, with the consent of the 'host government,' and in international waters outside of Thailand, but each time, they ran into political obstacles because of the geographical

51 Joe Quirk and Patri Friedman, Seasteading: How Floating Nations Will Restore the Environment, Enrich the Poor, Cure the Sick, and Liberate Humanity from Politicians (Free Press, March 20, 2018), ISBN: 9781451699272.

implications of the nation-states surrounding them. In the case of French Polynesia, political changes halted the project, and in the case of Thailand, the Thai Navy declared them criminal and hunted them down. Another such example is Liberland, a self-proclaimed nation on the Danube River that has constant problems with the neighboring countries of Croatia and Serbia in order to be able to even access the land. This is the problem when trying to create governance alternatives with territories attached to them: you're always going to be at the mercy of the nation-states around you, and their changing policies. These are evolving stories, and hopefully some will work out over time, but it's undeniable that territory creates a liability and a central point of attack for any governance 2.0 project.

Expandatory Bureaucracies – the EU Parliament and the US Federal State

Initially, the US government was built on a decentralized model based on strong state autonomy, with the federal government mostly acting for foreign policy interests. However, the federal government has seen a strong increase of influence across states, thus weakening the autonomy of the individual states, hence the function of the ability of 'voting with your feet' for US citizens. In the European Union, the Schengen Zone was initially set up to allow free movement for European citizens within European countries, a function that remains largely intact today. The European Union was set up for coordination purposes, such as making the movement between nations easier by aligning education systems for equal credits in other member states, healthcare systems, and so on. If it would have remained at that limit, it would have increased competition between member states through easily allowing residents the right to move wherever they wished, similar to how the US state systems were initially planned. However, the bureaucracy rapidly ballooned and created extra sets of regulations in every sector, making it a regulatory and tax burden on member states rather than increasing competition between member states. Bureaucrats tend to want to keep their work and increase the size of their departments. They also attract other bureaucrats with that same mindset. Therefore, working for more decentralized solutions within existing structures is often an unproductive undertaking.

Centralization – E-Gold

E-Gold was one of the early predecessors to Bitcoin. It was launched as a centralized and gold-backed digital currency in 1996 by Douglas Jackson and Barry Downey. It enabled users to make transactions in gold, silver, and other precious metals. The service saw substantial growth, peaking in 2006 with over US$2 billion in annual transactions. However, it faced legal challenges due to its use in illegal activities and lack of sufficient anti-money laundering measures. Following a US federal indictment in 2007 for money laundering and operating an unlicensed money transmitting business, the company and its directors entered a plea agreement in 2008. E-Gold ceased operations and implemented a claims process for account holders to access their funds. In the realm of digital currencies, the tale of e-gold stands as a cautionary example of the perils of centralization. Established as a digital gold currency, e-Gold, unlike decentralized currencies like Bitcoin, hinged on a centralized operational model with offices and storage locations, an incorporated company with shareholders, and centralized services. This centralized structure eventually became its Achilles' heel, making it susceptible to legal and regulatory interventions that decentralized networks often evade. It's likely tales like this that made Satoshi work so hard on building Bitcoin in a decentralized manner.

VISIONS OF TOMORROW: BITNATION AND BEYOND

When Bitnation first started off in July 2014, it was a novel idea, the first organization of its kind. The first few years of Bitnation were mostly dedicated to explaining the concept to the world and conducting various pilot experiments to showcase the technical viability of the idea. It was an uphill battle, but it paid off in terms of spreading the message. Within just a few years, the project had been covered by all mainstream media and regularly contacted by government officials, academics, and activists from around the world who wanted to learn more and experiment with the ideas and toolset. Several copycats popped up along the way, though none of them gained any traction or survived longer than a few months. The only successful attempt to date has been a major Silicon Valley investor publishing a book about the concept, essentially copying Bitnation's work, including its white papers, talks, and GitHub entries, which are widely available online and circulated in the "governance 2.0 community." People often overestimate the advantages of being the "first mover" in any given space, and I was certainly one of them. As one notable investor put it after a dinner conversation in Amsterdam: "The first-mover advantage is vastly overrated because the first mover needs to convince the market of the need for their service or product, and by the time they've done so, they might be too worn out to deliver that product effectively. Hence the second market mover will sweep in when the first one is exhausted and deliver it better. Such as Tesla and Edison or MySpace and Facebook." We scoffed at the theory at the time but do not anymore. It is less entertaining as a theory when you are on the receiving end.

However, the silver lining in this story is Moore's law, the technological acceleration that has taken place over the last ten years. In 2014, the Ethereum blockchain, with its entire ecosystem, had not yet been released. ChatGPT and other generative AI applications did not exist in the public domain. Cross-chain functionalities were not something anyone ever spoke about. Even though it is fair to say that we are still in the early stages of these technologies, a tremendous amount of progress has been made in the last ten years, making the Bitnation concept more feasible and requiring fewer resources than ever before. With ten years of experience, knowing the opportunities and pitfalls, no one is better placed to deliver a sustainable solution than the ones who pioneered it.

Speculative Scenarios: How Might These Evolve, and What Might Their Impact Be?

The game plan has shifted over the last decade. It's no longer about being first to market but about being the best at it, and that primarily comes down to the technology stack and surviving through the ups and downs of political and technological shifts. It's a marathon, not a race, to state an old cliche. When Bitnation started, people often asked, "Do you think you're going to succeed with replacing the nation-state system?" And I (Susanne) answered: "Of course, the shift will happen naturally anyways, because of things largely beyond our control such as global trade, global communications, and so on. It's just a matter of how it will change, not if. But it will probably take fifty years to do it." Well, that statement seems to hold its ground.

If you turn on the news, you might be momentarily confused by the constant bashing of nationalist news—from right-wing elections in Europe, pro-Trump crowds in the US, and anti-Semitic left-wing crowds across the world. However, if you take a bird's-eye view of these kinds of events, then it rapidly becomes clear that they are just the last dying screams of the Tyrannosaurus rex of the Westphalian system. They are potent today, and they might even become more potent over the next few years ahead. But are they likely to prevail long-term? Considering the accelerating change around us, they are more likely to be replaced.

What do we do about the trajectory of Westphalia then? Well, we do what the human species has done since the beginning of time. We move on. We adapt and we evolve. Large nations tend to have a great problem adapting and evolving with time, as their governments are often more complex and slow-moving, and often more diversified in their population. That has led to a tendency for large nations to impose more rules and restrictions on their citizens over time to curb any potential uprisings and other disturbances to the nation-state apparatus. Often, this includes enforcing stricter societal rules, harder punishments, and more invasive surveillance mechanisms. Countries presently engaging in such tactics include the United States, Saudi Arabia, Russia, China, and Iran. Those techniques might have worked temporarily in the past, but they don't have a record of accomplishment of holding indefinitely because eventually, citizens feel disconnected from the governance mechanism. They begin to feel like they are on the losing end of the equation, in which case it tends to implode, often in violent and bloody ways resulting in mayhem and bloodshed. The process of implosion is only likely to be accelerated in today's global world, where a lot of people regularly travel, have friends and family abroad they speak with on a regular basis, and have access to communication technologies. Sure, nations can impose censorship on the internet and so on, and it might have some short-term effect on curbing or changing the information, but at the end of the day, even in the strictest societies, VPNs are accessible, and information gets through at a speed greater than ever before.

If constant warfare against groups in society or foreign nations is the only thing that can keep people in any given nation united and working in favor of the government through a constant bombardment of misinformation campaigns, such as Russia and Iran are engaging in, then their countries will quickly become poor and generally dystopian, which is a slow death in itself, and will likely implode too, as all economies today are interconnected with the global economy. Hence, smart, and small nations are far more likely to survive in a post-Westphalian setting—nations that put their citizenry, economy, and technology first. We can already see how many small nations and city-states enjoy a higher GDP than their larger regional counterparts; some of these are nations without significant natural resources, such as Singapore, Denmark, Luxembourg, Andorra, Liechtenstein, Switzerland, and the Netherlands, for instance.

Dubai, the most famous emirate in the United Arab Emirates (UAE), has very little oil in comparison to some other emirates, like the capital of the UAE, Abu Dhabi. Instead, Dubai has created its wealth with real estate, friendly taxation and corporate incorporation laws, and tourism. Today, the revenue from oil amounts to less than 1 percent of its GDP.[6] This was mainly due to a strategy formulated when UAE discovered the oil; they decided that instead of fighting over the national oil resources, the emirates would instead peacefully compete against each other by having greater liberty to form their own legislations. It is one of the greatest governance success stories of the post-WWII era. Due to the resilience of smaller, agile nations and regions, they are likely to survive and, in some cases, thrive longer. This includes small nation-states, highly decentralized (regionally managed) nation-states, city-states, and voluntary communities.

This might sound a bit libertarian as a vision—the vision of voting with your feet or wallet—but it is not libertarian in the classically imagined libertarianism. Governments can come in all forms and shapes when it comes to structure, services, and economic models. However, it is libertarian in the sense that the core principle is the non-aggression principle (NAP), which is a moral framework that says all actions between individuals and groups should be consensual, regardless of what those interactions are. Hence, you can join a nation that practices a strict form of governance and surveillance if you choose, but no one can force you to do so based on where you were arbitrarily born. It is not a pacifist principle, as it allows for self-defense. The NAP is far from perfect, of course, and has many

52 Matthew A. Winkler, "Dubai's the Very Model of a Modern Mideast Economy," Bloomberg View, January 17, 2018, https://www.bloomberg.com/professional/blog/dubais-model-modern-mideast-economy/.

inconsistencies. Probably the greatest one is when it comes to handling children who are not yet self-sufficient and dependent on their parents. Like any principle, it should be applied where possible and not be taken as an absurd religion. However, Susanne believes that it is the only form of a minimalist moral guideline that could function on a global scale without intruding on personal views, religions, or culture.

The end view of network-states, however, is to decouple governance service provision entirely from geography. For instance, imagine if you could simply choose your governance service provider from your phone as easily as you choose your food delivery service or cab service. You might choose a collection of several different providers, one which combines all services in a user-friendly package, or no governance service provider at all. You might be a digital nomad who, when you enter a physical place, signs on to the terms and conditions of the city or region or commune for the time you stay there, just as you do when you stay in a hotel, or you might live somewhere permanently, perhaps accepting another form of terms and conditions as a result (like increased discounts on city services). Many cities might only choose to charge taxes on purchases, not on salaries, because of the non-feasibility of it. Virtual nations, aka network-states, might choose to charge per service, use an insurance model with a yearly fee, or go with a by-donation structure, depending on what sort of clientele they cater to. They might have partnerships with each other or with city-states to provide discounts. But regardless, anywhere you live in the world, you can simply choose your governance service provider and opt in or opt out of their services as easily as you opt in or out of any membership website.

The shift to the worldviews presented in this book will not happen overnight, and not all at once. Rather, prepare for a long and tumultuous road ahead, and build yourself up financially, mentally, and physically to be able to see it through to the end. There will also be many roadblocks along the way, especially for those who live in areas with volatile and oppressive governance, or in volatile parts of the world. As we've seen over the last decades, even seemingly peaceful and prosperous areas can turn volatile overnight. Hence, work cautiously, try to keep aware of privacy technologies if you ever need them, and be grateful for every small step towards what will hopefully be a freer, fairer and more prosperous future for all.

CHAPTER ELEVEN

ECHOES OF THE FUTURE

ALICE & BOB: THE LANGUAGE MODEL ERA

In the not-so-distant future, Alice and Bob, two research scientists, try to collaborate on an opt-in nation-building project across continents. They are blessed with advanced tools: the use of personal large language models (LLMs) that aid them in their thought and writing processes, access to a decentralized blockchain system, and an ability to convert their thoughts into machine-readable code. **Here's how the process works:**

Enhanced Understanding with LLMs: Alice shares her ideas with her LLM, which then elaborates on them, refines them, or suggests improvements. After this refining process, Alice sends her message to Bob. Bob's LLM then optimizes the content based on Bob's knowledge and preferences before presenting it to him, ensuring that the message is tailored to his understanding.

Direct Code Communication: When discussing complex algorithms or solutions, Alice and Bob convert their pseudocode ideas into real code snippets through their LLMs. The models automatically test the logic, refine the code, and then send it over. On the receiving end, the LLM translates the code back into pseudocode or natural language, depending on the recipient's preference at that moment.

Immutable Records on the Blockchain: Every significant decision, research result, or agreed-upon plan is recorded on the blockchain. This ensures that there is always a transparent, tamper-proof, chronological record of their collaboration. It reduces disputes and misunderstandings. Plus, if ever they want to patent a discovery or prevent a patent troll from hijacking their technology, the blockchain serves as immutable proof of their work's chronology.

Instant Translation and Cultural Contextualization: Alice, being from Japan, sometimes uses idioms or cultural references that Bob, from Brazil, might not immediately grasp. However, the LLMs instantly translate and provide cultural context, making their communication seamless. It bridges cultural gaps and enriches their collaboration.

Human-Machine Symbiosis: Over time, their LLMs begin to predict frequently asked questions, offering answers or code solutions even before being asked. It feels like an extension of Bob and Alice's own thoughts, but with the computing power and knowledge of the entire internet. Their brainstorming sessions become highly efficient, with the LLMs providing real-time data, potential pitfalls, and solutions based on historical precedents.

Public Feedback Loop: For certain ideas, Alice and Bob allow their discussions to be public on the blockchain so that peers around the world can provide input. With the help of their LLMs, Alice and Bob can assimilate this feedback, enhancing their research.

Emotion and Intent Recognition: The LLMs can analyze the emotional undertones of Alice and Bob's messages, ensuring that the tone is always constructive. If Alice is having a rough day, her LLM would inform Bob's LLM discreetly, asking for patience or suggesting a break.

In this possible future, Alice and Bob achieved a level of productivity and understanding that today might seem impossible. Their collaboration isn't just between two minds, but between two minds enhanced and expanded by the power of advanced AI, bridging the gap between human intuition and machine precision. The blockchain stands as a testament to their journey, an immutable timeline of discovery, debate, and progress.

In our book, we focus on Lucy, the AI companion, but it is important to note that due to Moore's law, personal large language models such as those used by Alice and Bob in the example eventually will emerge. These personal LLMs will have the potential to be decentralized and well trained. The key elements of the Alice and Bob example are LLMs, blockchain, and most importantly, human-machine symbiosis so that both Alice and Bob can use these tools to enhance their own productivity. The blockchain not only can keep immutable records but Layer-1 technologies such as Ethereum can store their immutable speech as smart contracts. Alice and Bob can speak in natural language and have that become the code of their agreement in the form of smart contracts.

"Sense and reference" is a crucial distinction in the philosophy of language introduced by Gottlob Frege. Here's a breakdown of the concepts with examples to help illustrate:

1. Sense (Sinn)

The "sense" of a term is the way in which it presents its referent. It relates to the meaning, mode of presentation, or the mental image it evokes. It's how a particular entity or object is thought of or perceived.

Example 1:

Term: "The morning star"

Sense: The star we see in the morning.

Example 2:

Term: "The current U.S. president"

Sense: The role or title of being the president of the United States at this moment. The sense doesn't tell you who the president is, but you understand the role or title.

2. Reference (Bedeutung)

The "reference" of a term is the actual entity or object to which the term points or refers. It's the "real-world" object or concept that the term signifies.

Example 1 (corresponding to the previous example):

Term: "The morning star"

Reference: The planet Venus.

Example 2 (corresponding to the previous example):

Term: "The current U.S. president"

Reference: Whoever is the U.S. president at the current time (e.g., Joe Biden as of the time of this publication).

Applying Sense and Reference:

Names with the Same Reference but Different Senses:

"The morning star" vs. "The evening star"

Sense: The former refers to the star seen in the morning, the latter to the star seen in the evening.

Reference for both: The planet Venus.

Here, the names refer to the same object (Venus) but present it in two different ways.

Descriptions That Change Reference Over Time:

"The current U.S. president"

Sense: The role or title of being the president of the U.S. at this moment.

Reference: This changes over time, from George Washington to Joe Biden and to future presidents.

The sense remains constant, but the reference changes as time progresses.

Terms with Sense but No Reference:

"The largest prime number"

Sense: The idea of a prime number that is larger than all other prime numbers.

Reference: There isn't one, because there is no largest prime number (primes are infinite).

This example illustrates that a term can have a clear sense (meaning) without having a corresponding real-world reference.

By distinguishing between the sense and reference of terms, Frege aimed to address complexities in language and meaning.[1] His distinction helps clarify issues like why two names can refer to the same thing while having different meanings, or why certain sentences can have clear meanings even if they don't refer to any actual entity in the world.

When we look at words or signs, a name, or a combination of words, we have the referent. This is distinctly different from the sense. The sense is the meaning, perception, mode of presentation, or description. The morning star and evening star both refer to the planet Venus, which means they have the same referent. The mode of presentation or sense is what is different, even though it's talking about the same thing. If we say that the morning star is the evening star, the perceiver would get valuable cognitive information.

Applying Frege's distinction to the Alice and Bob scenario brings us to the following:

1. Enhanced Understanding with LLMs:

Sense: The way Alice thinks about and articulates her ideas might be differ-

53 The Cambridge Companion to Frege, ed. by Tom Ricketts and Michael Potter (Cambridge University Press, 2010), ISBN: 9780521842066.

ent from Bob's perception or understanding.

Reference: The actual idea or concept Alice wants to convey remains constant. The LLMs help in ensuring the sense in which Bob receives the information aligns closely with Alice's intent, even if the presentation is adjusted for clarity.

2. Direct Code Communication:

Sense: The pseudocode or natural language descriptions Alice and Bob use might vary based on their individual styles or backgrounds.

Reference: The underlying algorithm or solution they are discussing.

3. Immutable Records on the Blockchain:

Sense: The reasons or motivations for recording a decision or result.

Reference: The actual decision or result that's recorded.

4. Instant Translation and Cultural Contextualization:

Sense: Alice's idioms or cultural references.

Reference: The actual meaning or idea Alice intends to convey.

5. Human-Machine Symbiosis:

Sense: The way Alice or Bob might articulate or think about a problem.

Reference: The core problem itself, which the LLM can assist in addressing.

6. Public Feedback Loop:

Sense: The varied interpretations and feedback from peers worldwide.

Reference: The core ideas or research topics Alice and Bob are discussing.

7. Emotion and Intent Recognition:

Sense: The emotional undertones or nuances in the messages exchanged.

Reference: The actual emotional state or intent of the sender.

In essence, Frege's distinction between sense and reference can be instrumental in understanding communication complexities. In the context of Alice and Bob, it helps to pinpoint how advanced tools can assist in ensuring that the sense in which information is received aligns closely with the original intent (or reference), despite the varied modes of presentation or understanding between different individuals.

The goal of using LLMs in this way, as illustrated by the above examples, is to bring Alice and Bob's minds closer together. The LLMs allow them to communicate more effectively, achieving a much deeper level of agreement between them than what we typically have today. The language model era is our current era, where human-machine symbiosis is being sought after by using what I will call "advanced auto-complete" AI. I do not consider LLMs to be a potential AGI because of their limitations, but they do act, in some ways, as advanced auto-complete because they can predict what you want to say in a way that is magical, and they can capture intent. LLMs can achieve this thanks to the training data, which enables the LLM to recognize patterns and contexts in text. The effectiveness of LLMs' ability to do this is currently limited by the ability of the user to input commands (often via text prompts), as well as the specific context.

LLMs can predict the next word in a sentence. A personal LLM does not think or have consciousness. Instead, it considers the preceding words implied from your input (context), and from this, it generates what it determines to be the most likely continuation. In this way, it is an advanced auto complete. The LLM achieves this by applying statistical mechanisms such as the transformer architecture, which this book will not further explore.

According to Dana's current understanding of these technological tools, Dana does not personally believe that LLMs will lead to AGI. Personal LLMs still have risks that could destabilize a society that is unprepared to make the necessary changes to provide safety guarantees. For example, there are risks where a user of LLMs can offensively abuse their powers in an irresponsible manner. To adapt to these potential abuses, defensive strategies are necessary. These strategies may use advanced technological tools, such as other forms of AI or LLMs of a specific type. Dana believes there needs to be a much greater investment from the private sector, as well as funding from the government, to promote tools to control the risks that are likely to emerge from the era of LLMs.

One of the major risks of the personal LLM era is the danger of misunderstandings. When the understandings are deeper, it could also mean that the misunderstandings are deeper. While Alice and Bob could be brought together in very deep ways, it could also mean that the divides between groups of Alices and Bobs become much deeper. Dana believes there is not yet enough knowledge about the impact of LLMs to determine the risk the problem of deep misunderstandings could bring.

The widespread adoption of personal LLMs could have profound effects on many facets of society. Let us dive into a speculative analysis of the potential benefits, risks, and a vision of a future where LLMs are as ubiquitous as today's smartphones or crypto wallets.

Potential Benefits of Personal LLMs:

Enhanced Productivity and Efficiency: Tasks that require information processing, research, drafting, coding, or other language-related work could be expedited. This might free up individuals to focus on tasks that require deeper human creativity, intuition, or emotional intelligence.

Education: LLMs could provide personalized learning experiences, adapting to each student's pace and style. They can assist in explaining complex topics, tutoring, or even helping with homework.

Language and Communication: Breaking down language barriers might become simpler. If each person has a personal translator that understands context deeply, global communication could be revolutionized.

Accessibility: For those with disabilities, LLMs can be invaluable. They could help in reading content out loud, understanding spoken language, assisting with writing, and more.

Decision Support: Everyday decisions, from shopping to more complex choices like investments, could be aided by LLMs that can quickly provide pros and cons or analyze vast amounts of data for insights.

Potential Risks of Personal LLMs:

Over-reliance: There's a danger of depending too heavily on LLMs. Critical thinking, deep reading, and personal reflection are essential skills that might diminish if people lean too much on machines.

Privacy Concerns: If personal LLMs operate over cloud platforms, there might be concerns about data privacy and the potential misuse of personal data.

Economic Displacement: Many jobs, especially those that revolve around information gathering, basic content creation, or data entry, might become obsolete, leading to economic and job market disruptions.

Misinformation and Manipulation: LLMs can generate false or misleading content. In the wrong hands, this capability could be used for propaganda, fraud, or scams.

Loss of Cultural Nuance: Over-reliance on LLMs for communication might lead to the erosion of linguistic and cultural nuances that enrich human interactions.

A Vision of a World with Ubiquitous LLMs:

Personal Assistants Everywhere: Much like how smartphones are a primary interface today, people might interact with their LLMs frequently for various tasks—from writing emails to cooking recipes, from learning a new topic to meditation guidance.

Collaborative Creativity: Artists, writers, and creators might use LLMs as collaborative tools to brainstorm, draft, or refine their works, ushering in a renaissance of co-created art between man and machine.

Decentralized Knowledge Access: With LLMs, access to knowledge might become even more decentralized. Rather than relying on specific platforms or websites, individuals could ask their LLMs, ensuring a more democratized access to information.

Continuous Learning: The boundary between formal education and daily life might blur. With constant access to an LLM, every curiosity, every question can lead to a deep dive, fostering a culture of lifelong learning.

Augmented Social Interactions: LLMs might help curate and facilitate social interactions, suggesting topics of common interest, breaking down language barriers, or even helping people understand and navigate complex emotions.

The lists of risks and benefits is not exhaustive, as it's impossible to map out all the possible risks of a new technology or a new societal paradigm. There are many potentially unknown and currently unknowable risks as well as benefits. But by providing some examples of potential risks and benefits, we can get a picture of what society could look like when personal LLMs are ubiquitous.

The Whispered Revolution

Let us look into the future, to the year 2045. In 2045, Earth is no longer dominated by large screens and the familiar tap-tap of keyboards. Instead, the world echoes with soft whispers—the conversations between humans and their personal LLMs, fondly nicknamed "Whisperers."

Maria, living in São Paulo, wakes up to her Whisperer's gentle voice giving her an overview of the day: news from around the world, a brief on the local weather, and personal updates like her niece's first spoken word in Madrid. As she prepares breakfast, she uses her Whisperer to translate a traditional Japanese omelet recipe, seamlessly switching between Portuguese and Japanese.

On her way to work, Maria notices a tourist struggling to ask for directions. She approaches and, with her Whisperer's assistance, provides directions in flawless Dutch, despite never having been to the Netherlands. The grateful tourist, amazed, remarks, "These Whisperers have turned the world into a global village, truly."

In downtown Nairobi, a group of students gathers around an old oak tree, each discussing the intricacies of quantum mechanics, Renaissance art, African folklore, and Andean music. Their Whisperers subtly provide insights, allowing debates to reach unprecedented depths. Far from isolating individuals, the Whisperers ignite a passion for collective learning.

Meanwhile, in a quiet corner of Stockholm, renowned author Axel Larsson, known for his riveting thrillers, confesses in an interview, "My last novel was a duet between my imagination and my Whisperer's expansive knowledge. It's like having a muse that understands the very pulse of global narratives."

However, it is not all rosy. In Shanghai, a support group called "Unplugged" meets weekly, advocating for time without Whisperers. They believed that while the Whisperers enrich, they also risk numbing innate human creativity. Their meetings, full of laughter, debates, and sometimes tears, are a testament to raw, unaided human emotion.

Back in São Paulo, as Maria settles into bed, she ponders a question that has been nagging her all day. "Whisperer," she murmurs, "what makes us human in this age of ubiquitous knowledge?"

The Whisperer responds, "While I can provide information, emotions, dreams, love, hope, and the intrinsic desire to connect make you human. My existence amplifies your reach, but the essence of humanity remains yours alone."

Maria smiles, comforted, and drifts off to sleep in a world that has found a balance between the whispers of machines and the heartbeat of human existence.

THE GROUP MIND AS LEGAL FICTION

The idea of combining LLMs like ChatGPT with knowledge and reasoning systems like TauChain or Cyc to form a collective "group mind" is intriguing. Such a union could manifest capabilities that are greater than the sum of its parts. The subsequent notion of this collective being recognized as a legal entity draws parallels with concepts of corporate personhood. Let's delve into this speculative scenario:

The Group Mind Network:

$$\text{Shared Cognition} = \sum_{i=1}^{n} \alpha_i \cdot (C_i + A_i) + \sum_{i=1}^{n} \sum_{\substack{j=1 \\ j \neq 1}}^{n} \beta_{ij} \cdot (C_i + A_i) \cdot (C_j + A_j) \quad (1)$$

By integrating LLMs' vast knowledge and language capabilities with the common-sense reasoning and logical processing of systems like Cyc or Tau, we'd get a network that can:

Absorb Information Rapidly: Reading and understanding vast amounts of text in real-time.

Reason and Make Decisions: Using logical frameworks to make informed decisions.

Interact Intelligently: Communicating its findings, decisions, or reasoning in human-like language.

This amalgamation would have capabilities that could rival complex human organizational structures in terms of decision-making, problem-solving, and information dissemination. It would essentially operate as a super-advanced, unified digital brain, a collective exocortex. Today, we have the concepts of corporate personhood, corporate responsibility, and the corporation as legal fiction. In our opinion in the future, we will have to see an update to the law in a similar manner, where decentralized networks acting as a unified person can be represented as such under the law in the same way that nation-states are considered sovereign entities.

Corporate personhood grants certain legal rights to corporations, treating them as individual entities. A group mind made up of atomic extended minds would need to be legally defined in a way that is distinct from a fully autonomous network. Earlier, we mentioned the term "decentralized autonomous corporation" (DAC), which fell out of favor and was replaced with the term "decentralized autonomous organization" (DAO) by the Ethereum community in 2016–2017. In these examples, you have the intention of a mostly or perhaps fully autonomous profit-making entity in a DAC while a DAO may, in fact, have another motive. In my current understanding, the symbiosis between humankind and machines, which I call "Cyborgization," is what will enable, at first, extended minds. These extended minds will be human beings who command groups of autonomous agents that act on their behalf according to their orders. Over time, these extended mind humans who, in a certain sense, we could call cyborgs, will increasingly have access to better AI, which can combine their will so that an individual mind becomes a group mind network.

A group mind network will have capabilities far beyond what a DAC or DAO was envisioned to have. A DAC is like a robot-run corporation where the CEO is encoded into the smart contract itself. But the DAC typically still has human stakeholders, human developers who upgrade or maintain it, and so on. A group mind network is different because this group mind can act in complete unison. And unlike a DAO, this group mind can leverage logical AI to understand common sense and knowledge input from the world. Because knowledge can be combined logically using knowledge representation languages, blockchains, ontologies, and so on, we can have a group mind that consumes knowledge as input and acts upon the knowledge according to the rules of its members.

The food source for a group mind network is knowledge. This knowledge is stored as a shared knowledge base. The ownership of such a network could be secured by a blockchain with tokens. This kind of network could make corporations as we understand them today look primitive because this kind of network combines AI into the organization. This group mind network is not restricted to corporations; it could be a way for humans to organize around philanthropy, religious institutions, or higher learning. In Bitnation, the Lucy AI conceptually was intended to evolve into a group mind network of sorts. This group mind network theoretically would allow all members to contribute their knowledge and their portion of understanding in a machine-readable way. Theoretically, this group mind network would determine code generation and software synthesis for Bitnation itself to continuously evolve in alignment with the members. By maintaining this alignment in a logical way, the intention is to have a level of AI safety that we do not see from generative approaches to AI.

The unique benefit of having a shared knowledge base is that a million or even billions of people can contribute their knowledge to make governance better. A lawyer with a deep understanding of international law could input their legal knowledge into their instance of Lucy and then contribute from their instance to the shared knowledge base of the group mind. This would instantaneously update all minds connected to the network with deeper legal understanding and awareness, which each independent mind or the group mind could utilize.

What is interesting about this is that when you have a group mind network that compounds in knowledge, the network, if given the right set of rules, can positively evolve at the rate and quality of new knowledge that is given as input.

This is different from LLMs, as every user cannot update the knowledge (at least not as of the time of writing), but with a shared knowledge base, this becomes possible. What also becomes possible is that each individual mind in the network can have their worldview, their moral system, but when combining into the group, the group can have some rules to only accept as members minds that are compatible. In this way, we can expect there would be many different group mind networks with various mindsets or manners of thinking because human beings do not universally agree on what is good and bad or how to measure good and bad choices. The benefit of this is that like minds will have a dramatically easier time finding each other than they do today with social media.

This fluidity of group adherence makes the group mind network approach perfectly suitable for non-geographical governance service providers, which people can opt in and opt out of as easily as they choose what social media website they use, without having to move to another geographic area.

THE PROBLEM OF IDENTIFYING OBJECTIVE TRUTH AND CONSENSUS TRUTH IN THE ERA OF LLMS

In our increasingly complex digital age, the very nature of "truth" and the way we come to discern it are under scrutiny. The emergence of information technologies, social media, and AI-driven platforms has accelerated the dispersion of both information and misinformation, creating a conundrum: How do we delineate what's true from what's perceived as true? This leads us to the crux of the issue—the distinction between "objective truth" and "consensus truth."

Objective truth refers to an absolute, unchanging fact that remains true regardless of belief or opinion. It's independent of individual and collective thought. For example, the statement "water boils at 100°C at one atmosphere of pressure" is an objective truth, verifiable by repeatable scientific experiments.

Figure 25: ***Objective truth: two scientists agreeing that the boiling temperature of water is 100 degrees celcius.***

Consensus truth is a form of truth that is generally accepted by a majority, often within a particular community, culture, or group. It's based on shared beliefs or perceptions, even if these beliefs don't align with objective facts. Historical narratives, certain cultural beliefs, or widely accepted social norms often fall into this category. They may change over time and vary between different communities or cultures.

THE PERSONAL SWARM: EXTENDING THE HUMAN MIND THROUGH AI AGENTS

In the ever-evolving landscape of artificial intelligence, the concept of a "personal swarm" stands out as a beacon of the future—a future where each individual is augmented by an ensemble of AI agents, a swarm dedicated to executing their will, thus extending the capabilities of the human mind. Imagine Alice and Bob, not as isolated individuals, but as nexus points for a myriad network of intelligent agents, each member of the swarm a reflection of their goals, desires, and thoughts.

The Swarm as an Exocortex

The personal swarm operates as an "exocortex," an externalized component of the human brain. Just as the neocortex allows for higher-order thinking and problem-solving, the exocortex enhances these cognitive processes through digital means. Each AI agent in the swarm specializes in different tasks—scheduling, communication, research, entertainment, health management, and more—working in concert to extend Alice and Bob's mental faculties.

The AI Companion as an Operating System

In earlier chapters, we discussed how Lucy could conceptually fill the function of an operating system for AI agents. In the future, this paradigm of using the AI companion or assistant as a "boss agent" to operate dozens or perhaps hundreds of agents could become the predominant way human beings interact with the computer. Just as browsers originally only had one tab and you had to either open a new browser or visit one website at a time, the average Internet user today has many open tabs and is browsing many websites simultaneously. For personal LLMs to scale into personal swarms, there will need to be some hardware efficiency breakthroughs to bring down the costs of either self-hosting personal LLMs or hosting them in the cloud.

Seamless Symbiosis

The interplay between the AI companion "boss agent" and the swarm will become seamless. This will create a symbiosis where the line between user intent and AI actions blur. The AI companion "boss agent" ensures tasks are executed efficiently, learning from each interaction to optimize future performance. It's a dynamic system, one which evolves along with the user, customizing its approach, as it gains insights into preferences and behaviors.

The Mind-Extending Capabilities of the Swarm

The capabilities of the personal swarm are vast:

Information Retrieval: Agents within the swarm can sift through terabytes of data to extract relevant information, extending the user's memory and knowledge.

Decision Support: By processing complex data and presenting it in an understandable form, the swarm aids in decision-making, thus enhancing the user's reasoning abilities.

Task Automation: Routine tasks are offloaded to the swarm, freeing the user to focus on creative and critical thinking.

Learning and Adaptation: The swarm can track the user's learning goals, curate educational content, and even facilitate the learning process, thus extending the user's intellectual growth.

The Role of the Boss Agent

The boss agent (Lucy, for example) is the conductor of this orchestra of AI agents, ensuring that the symphony of tasks is performed harmoniously. Lucy as a boss agent provides a user-friendly interface through which Alice and Bob interact with their swarm. With a focus on user experience, The boss agent interprets the nuanced desires of the user, managing the swarm with a level of sophistication that feels intuitive and almost humanlike.

The Future of Personal Swarms

As the technology matures, the personal swarms could become more integrated into daily life. The AI agents could anticipate needs, act proactively, and execute tasks with increasing autonomy, all under the vigilant management of the boss agent. The potential for personal growth and societal transformation is immense, as every individual gains the ability to harness a level of cognitive processing power that was once the realm of science fiction.

Central Node: At the center of the diagram, we have Lucy, represented as a larger node or circle. This signifies Lucy's role as the operating system and central hub of communication.

Surrounding Nodes: Around Lucy, there are multiple smaller nodes, each representing an individual AI agent in the personal swarm. These nodes might be various sizes to represent different functions or capacities of the agents.

Connections: Arrows or lines connect Lucy to each of the AI agents, illustrating the flow of information and commands. A two-way arrow between the user and Lucy indicates interaction and communication.

The Group Mind Decision Matrix

One inevitable question is how a group of thousands or millions of people would not only reach agreement but arrive at decisions in a mathematical way. One approach we may see, if these technologies do exist, is the implementation of decision matrices. The AI companion for each individual member of the network would have the task of facilitating group decision-making.

Here's a step-by-step process illustrating how this might work:

Step 1: Discussing Desired Results and Defining Criteria

The group mind network engages in a discussion facilitated by the AI, where the desired outcomes and features of the software are debated and defined.

Using natural language processing, the AI agent extracts key elements from the discussion to form a list of potential software specifications.

Step 2: Formalizing Criteria

The AI translates the extracted discussion points into formal criteria that can be quantitatively evaluated. It may ask for clarification if certain points are ambiguous.

Step 3: Weighting Criteria

The AI proposes a set of weights for each criterion based on the emphasis they received in the discussion or past decisions in similar contexts. These weights are presented to the group for approval or adjustment.

Step 4: Scoring Specifications

The group members use a dedicated interface to score each proposed specification against the formalized criteria. This scoring can be iterative, with the AI providing real-time feedback on the impact of different scores.

Step 5: Computing Decision Matrix

The AI applies the decision matrix algorithm, multiplying scores by the agreed weights and summing the results to get a total score for each specification.

Step 6: Synthesizing Program

Once there is agreement on the specifications, the AI uses program synthesis to translate the formal specifications into executable code.

Step 8: Iterating for Consensus

If there's no initial consensus, the AI can facilitate an iterative process, adjusting weights and scores based on further discussion until a satisfactory level of agreement is reached.

Group Decision Process via AI-Facilitated Decision Matrix

$$\text{Score}_i = \sum_{j=1}^{n} w_j \cdot s_{ij}$$

Here's what each symbol represents:

Scorei: Total weighted score for option i.

wj: Weight of criterion j.

sij: Score of option i for criterion j.

n: Total number of criteria.

These scores or point systems can be backed by tokens redeemable for computation resources. In this way, the software specification with the highest scores would be software with the most computation resources subscribed to it. This is, of course, entirely hypothetical, depending on such technology being invented, but the mathematics of the decision matrix process will work even in low technology with no AI. In a high-tech AI use case, the decision matrix would be a way for humans to quantify their priorities in a manner which is entirely computable should they choose to use their AI agents in that fashion.

It is important to understand that the simple decision matrix process explained in this book is only one of many quantitative decision-making processes a group can use to arrive at decisions that rationally align with their preferences. The decision matrix is simply one of the more well-known and easier-to-explain methods. For example, while an AI can theoretically synthesize a program, it's still critical that the user gives the instructions that best align with their preferences. To create a better output, the quality of the input must be as high as possible, which, for a computer, can mean highly quantitative speech compared to what most users are accustomed to.

GOVERNANCE ON A CHIP?

If the assumptions are that Moore's law will continue, and that cyborgization will continue, then at some point in the future, it will become feasible for governance to be at the chip level. The possibilities and risks of this have been discussed in academic papers, and one of the major risks is the cybersecurity risk. In the paper titled "*Securing the exocortex: A twenty-first century cybernetics challenge,*" written by Tamara Bonaci, Jeffrey Herron, Charlie Matlack, and Howard Jay Chizeck in 2014, the authors illustrate how difficult it would be to secure an exocortex.[2] For example, NeuraLink[3] currently has at least three hundred employees working on building a brain-to-computer interface, but for all but the disabled, how many will volunteer to inject a chip into their skull? The privacy and safety risks of doing that are unknown, and the potential that something could go wrong is high. On the other hand, smartphones, pins, watches and other wearables might offer a degree of convenience without the same physical risks.

How might all of this evolve? Below is an illustration in the form of several short stories:

Year 2030: The Breakthrough

Advancements in quantum computing and AI, propelled by Moore's law, lead to the development of the first prototype of the governance chip. This chip, no larger than a grain of rice, can process vast amounts of data, learning and adapting to new governance challenges. It is tested in a small, tech-savvy city-state, which sees unprecedented levels of civic engagement and efficient decision-making.

54 Tamara Bonaci, Jeffrey Herron, Charlie Matlack, and Howard Jay Chizeck, "Securing the Exocortex: A Twenty-First Century Cybernetics Challenge," 2014.

55 Neuralink, https://neuralink.com/.

Year 2040: The Expansion

The success in the city-state becomes a global phenomenon. Countries, intrigued by the promise of reduced bureaucracy and increased transparency, start adopting the governance chip. The chip's AI has been programmed to understand cultural and societal nuances, making it adaptable to different regions. International agreements are established for data sharing and security standards, laying the foundation for a global network of digital governance.

Year 2050: The New Era

By 2050, the governance chip has evolved beyond recognition. Now embedded in everyday devices, it facilitates real-time citizen feedback and participatory decision-making. Governments are more agile, responding swiftly to public needs and global crises with data-driven strategies. The chip's AI, bound by ethical protocols, ensures fairness and impartiality, transforming how justice and legislation are perceived.

The Turning Point

A global environmental crisis tests the system. The governance chips, now interconnected globally, collectively analyzes environmental data, predicting catastrophic outcomes. For the first time in history, a unified global response is initiated, harnessing the collective resources and efforts of nations, guided by the impartial and data-driven recommendations of the governance chips.

These examples illustrate a possible evolutionary trajectory and are very speculative. For now, the governance chip is science fiction in terms of feasibility. By 2030, 2040, or 2050, it might increasingly become feasible, and if it is feasible, it's a technological evolution that could divide humanity.

Hugo de Garis, a researcher in the field of artificial intelligence, is best known for his work on what he terms "artilects" (artificial intellects), and his book *The Artilect War: Cosmists vs. Terrans.*[4] The book delves into a speculative future

56 Prof. Dr. Hugo de Garis, The Artilect War: Cosmists vs. Terrans: A Bitter Controversy Concerning Whether Humanity Should Build Godlike Massively Intelligent Machines (Etc Pubns, 2005), ISBN: 0882801546.

centered around the development of highly intelligent machines and the societal and ethical dilemmas that arise from such advancements.

De Garis argues that the future development of highly advanced AI, or artilects, will lead to a profound ideological divide among humans. He categorizes people into two groups based on their stance toward these artilects: the "Cosmists" and the "Terrans." Cosmists are those who support and advocate for the creation and development of super-intelligent machines, seeing them as a natural and beneficial progression of human technology and intelligence. They believe that artilects could help solve complex problems and potentially aid in understanding the universe at levels beyond human comprehension.

On the other hand, the Terrans are deeply opposed to the development of artilects, driven by fears that these entities could surpass human intelligence and potentially pose existential threats to humanity. Terrans argue for strict limitations or outright bans on the development of such advanced AI to prevent what they perceive as a potential loss of human control and sovereignty.

De Garis predicts that this ideological split could lead to a global conflict, which he calls the "Artilect War." This war, he speculates, wouldn't be just a political or ideological conflict but could escalate into a physical, potentially catastrophic battle over the future direction of AI development. His views highlight the potential risks and ethical considerations in advancing AI technology, emphasizing the need for careful thought and planning in how humanity approaches these powerful technologies.

"The Artilect War" is a thought-provoking exploration of potential future scenarios in AI development. It serves as a warning and a philosophical inquiry into the future of technology, intelligence, and the very nature of human existence in a world where artificial intelligences might surpass our own.

The middle path De Garis mentions in his book is the path of cyborgization. This is where some humans will choose to merge with and co-evolve in symbiosis with the AI rather than fight against it or build a post-human AI. Elon Musk elaborated on a term he called "speciesism," where the default stance is that humanity comes first, is considered special, and is at the summit of the metaphorical "food chain." But this debate is far more nuanced than what has been talked about in public. We really need to ask some deep philosophical questions about whether we should design AI to be human-like, with a sense of self, with the possibility of feelings or human-like rather than tool-specific qualifications. If an AI has human-like qualities, and it's also smarter than us, and we decide we are the kings of the universe by default, could this lead to a possible disaster? I think it's ultimately up to us, and how we design the AI, and whether or not we set it up to be a competitor or a tool.

In 2001: A Space Odyssey, the AI system HAL was necessary to guide the ship. In reality I don't think humans can travel far in space or escape planet earth without AI and most likely AGI or even superintelligence. Why? As isolated individuals, none of us are smart enough to become interplanetary. It's only when we can combine our minds into a super mind, and combine our computers into a supercomputer, and align both, that we have a shot at spreading life at the cosmic scale.

It's unknown at this time if the cyborgization approach or middle path is the best option. Our opinion is that it's a safer option than to risk AI evolving completely independently of humanity, but it's not clear exactly what humanity will evolve into when merged with AI. The speciesists and "Terrans" have a strong argument on AI safety and existential risks, and their argument becomes stronger as statistically measurable risks increase. If we had AGI, this debate would be a lot louder than it is now, where we have a public perception of AGI that exceeds reality. If we had actual AGI in the world, the majority of people globally would be impacted, and it would become the number-one topic of political discussion. The nightmare scenario presented by Hugo de Garis in his book is not beyond the realm of possibility. It might not be feasible today, but we can already feel the winds blowing, and the seeds of psycho-political fault lines forming.

This book, like the writing of Hugo de Garis before it, is meant to inspire a more nuanced discussion of AI, of governance, and of the politics surrounding the evolution of both. It's also about discussing how both AI and governance can potentially merge in a symbiotic way that we are calling "cyborgization of governance." In the near term, this cyborgization of governance can improve and augment the ability to govern in the same way having a library, glasses, computers, or a smartphone once did. In the more distant future, it may lead to a more inclusive world of network-states that can even replace the current geopolitically defined nation-states and international organizations.

ABOUT DANA AND SUSANNE

DANA EDWARDS

Dana Edwards stands at the intersection of technology and innovation. Holding a bachelor's degree in philosophy from the University of Massachusetts Boston and a master's degree in cybersecurity from the prestigious University of Maryland Global Campus, Dana's expertise is solidly anchored in academic rigor. Yet, it's the transformative power of literature and pioneering projects that have truly shaped Dana's perspective.

Books like James Altucher's Choose Yourself, which champions the ethos of individual resilience and creativity, and David Ronfeldt's Cyberocracy Is Coming, exploring the futuristic blueprint of digital governance, have been cornerstones of Dana's intellectual journey. The aspirational vision offered by James S. Albus in Path to a Better World has also been instrumental in shaping Dana's thoughts about a technologically integrated future.

But it isn't just literature that inspires Dana. The revolutionary concept of Bitcoin introduced by the enigmatic Satoshi Nakamoto and the avant-garde Zerostate Project have both fueled Dana's passion, cementing a belief in the potential of technology to redefine boundaries and empower individuals.

In this book, Dana not only shares expert insights but also invites readers into a world where technology and human potential harmoniously intersect. Dana's intention in writing this book is to promote ideas that can encourage the building of a freer, more secure, and happier society.

SUSANNE TEMPELHOF, INVENTOR OF NETWORK-STATES

Susanne Tempelhof is an international entrepreneur, award-winning innovator, frontier governance and technology researcher. With a commitment to pushing boundaries, she has redefined traditional notions of governance and identity, leaving a lasting impact on the global stage.

As a second-generation immigrant with parents from two different countries, the daughter of an undocumented and stateless refugee, she intimately understands the profound search for nationhood and identity in a world that often fails to provide a suitable framework. In 2004, she started writing about the concept that later would be known as Bitnation, but which she at the time called the significantly less catchy name: "Non-Geographical Governance Service Providers." The idea was that anyone, independent of the location of their birth or residency, could choose their governance service provider on an international market. She believed it would have to be set up as a sort of international insurance company and require hundreds of millions in funding. Thus, she parked that idea for a few years and decided to build experience, connections, and wealth within traditional government structures before launching her own governance solution.

In 2008, Susanne founded Wise in Afghanistan, conducting social science research on local tribal and governance structures, primarily for the US Department of Defense. This led to the acquisition of contracts worth more than US$7M and the successful sale of Wise to a US contracting firm in 2011. Onward to the Arab Spring, she established Shabakat Corporation in 2011, supporting grassroots movements during the civil war in Libya and the revolution in Egypt. Recognizing the limitations of conventional 'nation-building' approaches, Susanne embarked on a quest for alternative models of governance once again.

She stumbled across Bitcoin at the end of 2011 and was blown away by its premise of a decentralized, global currency. By the beginning of 2013, she had left the field of government contracting entirely and set off to write a book called The Googlement about the idea. While doing research for the book, in the beginning of 2014, she came across the Ethereum white paper and the concept of smart contracts and realized this was the key to launch her idea, with only a small fraction of the funds needed by her previous estimates a decade earlier. Excited by the emerging technologies, she decided to put the book on hold, launch her post-nation-state concept, and write a book about it afterward instead.

Hence, in 2014, she made history by launching Bitnation, the world's first network-state, powered by the blockchain. Bitnation revolutionized the concept of governance with pioneering initiatives such as the World Citizen ID, Marriage and Birth Certificate, and Land Titles, sparking a global conversation on decentralized governance. In partnership with Estonia's e-Residency program in 2015, Susanne introduced a public notary service, showcasing the global potential of blockchain in delivering public services. The same year, she launched the Bitnation Refugee Emergency ID Program, which received the prestigious UNESCO NetExplo Award in 2017 for its humanitarian impact during the European refugee crisis. Building on these achievements, she developed a global network of embassies and ambassadors, introduced a peer-to-peer jurisdiction accessible through Android and Apple apps, raised more than $2 million, and launched a token worth over $300 million at the time of launch.

In 2017, WIRED magazine acknowledged her contributions as one of the "Ten Innovators Fighting for Your Right to Privacy Online." She's a regular media contributor and keynote speaker at conferences, having appeared in publications such as The New York Times, Forbes, WIRED, VICE, TechCrunch, The Economist, BBC Radio 4, The Telegraph, CNN,

Bloomberg, Canal+, Wall Street Journal, Foreign Policy, NASDAQ.COM, The Atlantic, The Independent to name a few, and has spoken at events such as TEDx, Global Leaders Forum, the University of Cambridge, the London School of Economics, the EU Parliament, amongst many others.

With a vast international background, Susanne has traversed over sixty-five countries and lived and worked in diverse regions, including France, the United States, the United Kingdom, the Netherlands, Sweden, Egypt, Libya, Ghana, Brazil, China, and Afghanistan. Fluent in English, French, Swedish, Latin, and Portuguese, she embraces the richness of global perspectives.

BIBLIOGRAPHY

Asimov, Isaac. I, Robot. Gnome Press, 1950. ISBN: 978-0553294385.

Asor, Ohad. "Tau Whitepaper." Last modified March 1, 2023. Accessed March 18, 2024. https://tau.net/whitepaper_community_draft.pdf.

Bogdanov, Dan. "Foundations and properties of Shamir's secret sharing scheme research seminar in cryptography." University of Tartu, Institute of Computer Science 1 (2007).

Bonaci, Tamara, Jeffrey Herron, Charlie Matlack, and Howard Jay Chizeck. "Securing the Exocortex: A Twenty-First Century Cybernetics Challenge." 2014.

Chung, Frank. "'Bury them alive!': White South Africans fear for their future as horrific farm attacks escalate." News.com.au, 2017. https://www.news.com.au/finance/economy/world-economy/bury-them-alive-white-south-africans-fear-for-their-future-as-horrific-farm-attacks-escalate/news-story/3a63389a1b0066b6b0b77522c06d6476.

Clark, Andy, and David Chalmers. "The Extended Mind." Analysis 58, no. 1 (January 1998): 7–19. DOI: https://doi.org/10.1093/analys/58.1.7.

Creemers, Rogier. "China's Social Credit System: A Big-Data Enabled Approach to Market Regulation with Broad Implications for Doing Business in China." The China Quarterly 234, no. 4 (2018): 805–824.
Cycorp. https://cyc.com/.

Dalrymple, William. TheAnarchy: The East India Company, Corporate Violence, and the Pillage of an Empire. Bloomsbury Publishing, 2019. ISBN 978-1635573954.

De Garis, Hugo. The Artilect War: Cosmists vs. Terrans: A Bitter Controversy Concerning Whether Humanity Should Build Godlike Massively Intelligent Machines. Etc Pubns, 2005. ISBN: 0882801546.

Dunbar, Robin. How Many Friends Does One Person Need?: Dunbar's Number and Other Evolutionary Quirks. Faber & Faber, 2010. ISBN: 978-0571253432.

Emord, Jonathan W. Sharia Law: A Grisly Tale of Murder, Rape, and Incest. Creators Publishing, 2017. ISBN 978-1942557963.

Ford, Bryan. "Pseudonym Parties: An Offline Foundation for Online Accountability." March 27, 2007.

Ford, Martin. "AI and the Future of Work: Redefining Employment in the Age of Automation." Harvard Business Review, 2015.

Hanhimäki, Jussi M. The United Nations: A Very Short Introduction. Oxford University Press, 2008. ISBN 978-0199233163.

Hidalgo, Cesar A. and Ricardo Hausmann, eds. Economic Inequality: New Methods for New Trends. Palgrave Macmillan, 2018. ISBN 978-3319904241.

"Jasmine Revolution." Encyclopedia Britannica. December 10, 2023. https://www.britannica.com/event/Jasmine-Revolution.
Joyce, Christopher. "Palau Could Disappear In The Pacific Ocean Within A Century." NPR, 2017.

Karov, Georgi. The Force in Bitcoin (November 11 2021). ISBN: 979-8670620871.

Lafleur, Jean-Michel. "The Belgian Political Crisis of 2010-2011: An Analysis of the Causes and Consequences." West European Politics (2013). DOI:10.1080/01402382.2013.761222.

Ledger. https://www.ledger.com/.
Leroux, Robert, and David M. Hart. "Maurice Block on 'Decentralization' (1863)." In French Liberalism in the 19th Century, 1st ed. 2012.

Lessig, Lawrence. Code: And Other Laws of Cyberspace. New York: Basic Books, 1999.

Levine, Rick, Christopher Locke, Doc Searls, and David Weinberger, The Cluetrain Manifesto: The End of Business as Usual. Basic Books, 2001. ISBN: 978-0465024094.
Licklider, J. C. R. "Man-Computer Symbiosis." IRE Transactions on Human Factors in Electronics 1, no. 1 (March 1960): 4–11. DOI: 10.1109/THFE2.1960.4503259.

Lindenfors, Patrik, Andreas Wartel, and Johan Lind. "'Dunbar's Number' Deconstructed." Biology Letters 17 (2021): 20210158. https://doi.org/10.1098/rsbl.2021.0158.

Lu, Fangwen, Weizeng Sun, and Jianfeng Wu. "Special Economic Zones and Human Capital Investment: 30 Years of Evidence from China." American Economic Journal: Economic Policy 15, no. 3 (2023): 35-64.
Masson-Forsythe, Heather. "NSF gave Duolingo its wings!" National Science Foundation, December 22, 2022. Accessed 05/03/2024. https://new.nsf.gov/science-matters/nsf-gave-duolingo-its-wings#:~:text=In%20the%20U.S.%2C%20more%20people,%2C%20Duo%2C%20needs%20little%20introduction.

Maute, Jutta. Hyperinflation, Currency Board, and Bust. Peter Lang International Academic Publishers, 2018. DOI: 10.3726/b13965, ISBN: 9783631754474.

Neuralink, https://neuralink.com/.

Nygren, Johan. "Pseudonym Pairs: A foundation for proof-of-personhood in the web 3.0 jurisdiction." 2019.

Osiander, Andreas. The Westphalian System. London: Routledge, 2001. ISBN: 978-0415258813.

Palantir Technologies. https://www.palantir.com.

Parenti, Alessandro, and Marco Billi. "Building Trust in Smart Legal Contracts." In Transparency or Opacity, pp. 95–114. Nomos Verlagsgesellschaft mbH & Co. KG, 2023.

Parsons, C. "The Liar Paradox." Journal of Philosophical Logic 3 (1974): 381–412, https://doi.org/10.1007/BF00257482.

Quirk, Joe, and Patri Friedman. Seasteading: How Floating Nations Will Restore the Environment, Enrich the Poor, Cure the Sick, and Liberate Humanity from Politicians. Free Press, March 20, 2018. ISBN: 9781451699272.

"Research," OpenAI, https://openai.com/gpt-4.

Ricketts, Tom, and Michael Potter, eds. The Cambridge Companion to Frege. Cambridge University Press, 2010. ISBN: 9780521842066.

"Rohingya." Human Rights Watch. Accessed 5 March 2024. https://www.hrw.org/tag/rohingya#:~:text=In%20Myanmar%2C%20about%20600%2C000%20Rohingya,coup%20and%20brutal%20nationwide%20crackdown.

Searls, Doc. The Intention Economy: When Customers Take Charge. Harvard Business Review Press, 2012. ISBN: 978-1422158524.

Sherry, Benjamin. "What Is the Genesis Block in Bitcoin Terms?" Investopedia, updated November 05, 2023. https://www.investopedia.com/news/what-genesis-block-bitcoin-terms/

Sidewalk Labs. https://www.sidewalklabs.com/.

Skinner, B. F. Science and Human Behavior. Free Press, 1953. ISBN: 978-0029290408.

Stephen, Chris. "Libyan rebels struggle to set up alternative government." The Guardian, April 3, 2011.

Stern, David. "Ukraine's new president: How comedian Volodymyr Zelensky beat corruption." BBC News, 2019.

Sullivan, Clare, and Eric Burger. "E-residency and Blockchain." Computer Law & Security Review 33, no. 4 (2017): 470-481. ISSN 0267-3649. https://doi.org/10.1016/j.clsr.2017.03.016.

Tarkowski Tempelhof, Susanne. Bitnation Whitepaper. 2016. https://github.com/Bit-Nation/Pangea-Docs/blob/master/Bitnation%20Whitepaper%202014.pdf.
"The Chicago Question: What would Milton Friedman do now?" The Economist Magazine, 2012. https://www.economist.com/finance-and-economics/2012/07/28/the-chicago-question.
"The Netexplo Forum celebrated its 10th edition." UNESCO. Accessed 05/03/2024. https://www.unesco.org/en/articles/netexplo-forum-celebrated-its-10th-edition.

Widdows, Dominic, Kirsty Kitto, and Trevor Cohen. "Quantum Mathematics in Artificial Intelligence." Journal of Artificial Intelligence Research 72 (2021): 1307-1341- https://doi.org/10.48550/arXiv.2101.04255.

Winkler, Matthew A. "Dubai's the Very Model of a Modern Mideast Economy." Bloomberg View (January 17, 2018). https://www.bloomberg.com/professional/blog/dubais-model-modern-mideast-economy/.

Worldcoin. https://worldcoin.org/.

Wyatt, Jocelyn. "How climate change is changing the definition of 'refugee.'" OECD. 2023. https://www.oecd-forum.org/posts/how-climate-change-is-changing-the-definition-of-refugee#:~:text=The%20report%20also%20urges%20us,response%20to%20the%20climate%20crisis.

Zelazny, Roger. Lord of Light. 1967.
Zhang, Lena L. "Behind the 'Great Firewall': Decoding China's Internet Media Policies from the Inside." Volume 12, Issue 3. DOI: 10.1177/1354856506067201.

Printed in the USA
CPSIA information can be obtained
at www.ICGtesting.com
CBHW072245060724
11233CB00016B/1325